3 4028 06619 3880
HARRIS COUNTY PUBLIC LIBRARY

632.509 Eve
Everitt, J. H.
Weeds in South Texas and
    northern Mexico : a guide
    to identification
                        $19.95
                  ocm85692857
                  01/25/2008

We
and

W0082236

WITHDRAWN

Weeds in South Texas and Northern Mexico

A GUIDE TO IDENTIFICATION

James H. Everitt,
Robert I. Lonard, and
Christopher R. Little

TEXAS TECH UNIVERSITY PRESS

Copyright © 2007 Texas Tech University Press

Unless otherwise stated, photographs copyright © 2007 of the authors.
All rights reserved. No portion of this book may be reproduced in any form or by any means, including electronic storage and retrieval systems, except by explicit prior written permission of the publisher. Brief passages excerpted for review and critical purposes are excepted.

This book is typeset in ITC Stone Serif Medium.
The paper used in this book meets the minimum requirements
of ANSI/NISO Z39.48–1992 (R1997). ∞

Designed by Kaelin Chappell Broaddus

Library of Congress Cataloging-in-Publication Data
Everitt, J. H.
    Weeds in South Texas and northern Mexico : a guide to identification /
James H. Everitt, Robert I. Lonard, and Christopher R. Little.
      p. cm.
      Summary: "Identification guide to the 188 most common species of weedy plants
in South Texas and Northern Mexico. Presents information to identify the plants,
including a color photograph of each, as well as general comments about the habits
of the plants, their uses, and their possible toxicity"—Provided by publisher.
      Includes bibliographical references and index.
      ISBN-13: 978-0-89672-614-7 (hardcover : alk. paper)
      ISBN-10: 0-89672-614-2 (hardcover : alk. paper) 1. Weeds—Texas, South—Identification.
2. Weeds—Mexico, North—Identification. I. Lonard, Robert I. II. Little, Christopher R. III. Title.
      SB612.T4E94 2007
      632'.509764—dc22
                                                                      2007008279

Printed in the United States of America
07 08 09 10 11 12 13 14 15 / 9 8 7 6 5 4 3 2 1

Texas Tech University Press
Box 41037
Lubbock, Texas 79409–1037 USA
800.832.4042
ttup@ttu.edu
www.ttup.ttu.edu

CONTENTS

PREFACE

This publication contains color photographs, family names, common names, scientific names, general descriptions, and ecological characteristics of weedy plants that occur in South Texas and northern Mexico. It covers broad-leaved herbaceous species, grasslike plants, and grasses. The complete name history (synonymy) is not included in this work.

Although this publication focuses on plants that occur in South Texas and northern Mexico, the extensive ranges of many of the represented species make it a useful reference for weeds in other areas of Texas and the southwestern United States.

Some 187 species identified as weedy vascular plants in the South Texas area are included here, encompassing 144 genera and 45 families. These include 1 species of fern, 142 species of dicots, and 44 species of monocots. Of the dicot species, 111 are native and 31 are introduced. Of the monocot species, 20 are native and 24 are introduced. Of these 187 species, 56 are introduced. A native plant is indigenous to an area and grows without cultivation, whereas an introduced plant was brought in from another region and reproduces itself in the area. Voucher specimens for most of the plants included here are on file in the University of Texas–Pan American Herbarium (PAUH).

This publication will be useful to farmers and farm managers, agricultural consultants, ranchers, natural resource managers, scientists, and anyone interested in the flora of South Texas and northern Mexico. It will enable them to identify weedy species of this region.

The authors thank Mario Alaniz for his help in obtaining many of the photographs and for setting up digital files. Thanks are also extended to David Lonard, Rene Martinez, Daniel Flores, and Veronica Guzman for providing photographs of some of the plants and to Nick Hoelscher (Valley Garden Center, McAllen) for his assistance in locating several species. We thank Alfred Richardson, the University of Texas–Brownsville, for reviewing the manuscript and for providing photographs of some of the plants. We also express our gratitude to A. Michael Powell, Sul Ross State University; Peter Dotray, Texas Tech University; and an anonymous reviewer for commenting on the manuscript.

Weeds in South Texas
and Northern Mexico

INTRODUCTION

The primary intention of this work is to enable one to identify the weedy plants of South Texas, including the vast agricultural area of the Lower Rio Grande Valley (LRGV). However, the guide should be beneficial for identification of weeds for many other areas of Texas and northern Mexico. We hope that the color photographs, descriptions, and comments about specific weedy species will assist the general public and researchers in their efforts to identify and manage weeds in urban and rural landscapes of this subtropical area.

South Texas, and especially the LRGV, is one of the fastest-growing regions in the United States. The most recent census indicated a steadily growing population that increased from approximately 875,000 in 1990 to 1.2 million in 2000 (http://www.census.gov). In addition, 150,000 to 160,000 visitors spend the winter months in Starr, Hidalgo, Cameron, and Willacy counties. Census figures from Mexico catalog populations of 3.8 million and 2.7 million in the states of Nuevo León and Tamaulipas, respectively, that border the LRGV of Texas (http://www.ineg.gob.mx/inegi/default.asp). The population of Mexico, like that of Texas, is concentrated within 240 km (150 mi) of the Rio Grande.

As a result of a burgeoning population with its ensuing habitat destruction and "development," new environments have been created for both native and exotic weedy species. No treatment of the diverse weedy flora of the region exists. Many weeds in the region are native species, others are listed as globally significant weed problems, some are toxic or federally listed noxious species (Appendix 1), and others are invasive species that have an impact on the native vegetation of South Texas and northern Mexico, including rangelands and wetlands. Further evidence for the necessity of an identification guide has been noted by numerous requests for assistance in the identification of weeds by farmers, agribusiness agents, nursery professionals, ranch managers, landscape architects, researchers, and lay people who face problems in weed control and management.

The economy of the LRGV is centered on agriculture, trade, manufacturing, tourism, and oil and gas production. Agribusiness includes crop production (sorghum [*Sorghum bicolor*], cotton [*Gossypium hirsutum*], onions [*Allium cepa*], and vegetables), citrus production, cattle grazing, and forage production. Approximately 145,692 ha (360,000 acres) of sorghum, 84,987 ha (210,000 acres) of cotton, and 45,731 ha (113,000 acres) of onions were planted in 2005 in the LRGV (http://www.agr.state.tx.us). Data from the 2002 agricultural census (http://nass.usda.gov/tx/index.htm) indicate that approximately 12,141 ha (30,000 acres) are used for citrus production and 22,663 ha (56,000 acres) for forage. About 14,569 ha (36,000 acres) of vegetable crops are planted in Hidalgo County annually.

This guide is designed for a broad audience. Some users may have some botanical knowledge, but others may have only a cursory knowledge about plants. We have used technical terms where necessary to enhance clarity and conciseness, and we have defined those terms in everyday language. We have usually not reported measurements of plant structures.

Taxonomy and Nomenclature

The sequence of families, genera, and species is arranged alphabetically within the classes Polypodiopsida (ferns), Magnoliopsida (dicots), and Liliiopsida (monocots). Nomenclature follows Jones and Wipff (2003) for scientific names and the Sub-committee on Standardization of Common and Botanical Names of Weeds (1966) for common names. One or more common names are followed by a scientific name, and the Latinized scientific name is followed by the name of the author who first validly published a description of the species. His or her name is placed in parentheses if the taxonomic rank has been changed, and the author responsible for the taxonomic change is added. In this book a second name or synonym (Syn.) occasionally follows.

What Is a Weed?

A weed is a plant that is unwanted because of its undesirable features or a plant that is growing out of place. For example, bermudagrass (*Cynodon dactylon*) is a common introduced lawn grass in South Texas, but it may be a weed in farmland or in a flower bed. Under certain conditions it may even be toxic to livestock or cause hay fever in humans (Hart et al. 2003).

Weeds invade cultivated crops, gardens, and lawns. They have increased on rangelands where native perennial grasses have been removed or where native grass cover is too sparse to compete with aggressive weedy species, and they quickly become established along roadsides, construction sites, railroads, canal banks, power line rights-of-way, and dump sites. The most significant loss from pernicious weeds occurs through the depletion of soil moisture (Janick et al. 1981; Monaco, Weller, and Ashton 2002; Byrd 2003). Soil moisture is frequently the limiting factor in crop yields in the LRGV.

Weeds include broad-leaved herbaceous plants (dicots), grasses, shrubs, and trees, but weedy shrubs and trees are not included in this book. The interested reader may refer to Everitt, Drawe, and Lonard (2002) for information about these plants in South Texas.

What Is the Source of Weeds?

Many of the worst field and lawn weeds in the LRGV, including guineagrass (*Urochloa maxima*), buffelgrass (*Pennisetum ciliare*), bermudagrass, Kleberg blue-stem (*Dichanthium annulatum*), and Angleton bluestem (*D. aristatum*), have been introduced into the area from tropical or subtropical regions of the Old World. Therefore, these weeds are referred to as invasive species because they disrupt native South Texas ecosystems. A few weedy species (for example, London rocket [*Sisymbrium irio*]) that have been introduced from Europe are present in agricultural fields during the cooler winter months.

Other weeds, including Palmer amaranth (*Amaranthus palmeri*) and coastal sandbur (*Cenchrus spinifex*), are common native species that are widely distributed in North America. The status of a weedy species has been indicated under the "General Comments" section of each description as *native* or *introduced*.

How Do Weeds Spread?

In 1848, introduced weedy species constituted about 10 percent of the vascular plant species in the United States. However, they make up about 30 percent (5,000 species) of the total flora (17,000 species) in the country today (Pimentel et al. 2000). Invasive weeds must be managed intensively because they will dominate crop fields and native plant communities. Invasive species grow rapidly, reproduce by several means, and survive in a variety of temperature, light, and soil conditions. They are difficult to control and are almost impossible to eradicate.

Damage inflicted by invasive weedy species in the agricultural economy is estimated at $34 billion annually in the United States (Pimentel et al. 2000). Losses caused by weeds exceed losses caused by all other categories of agricultural pests (Anonymous 1971). Invasive species cover an area in North America about the size of California.

Fruits and seeds of weedy plants are often equipped with barbs, beards, claws, spines, or aerodynamic features that aid in dispersal. Weeds have spread quickly in the environment from many sources, including

- unclean crop seeds
- weeds on the land that have been allowed to produce seeds
- livestock feeds
- seeds in animal excrement
- farm machinery
- animal feet and body coats
- garden plants for transplanting, rhizomes, and seeds
- nursery stock, especially balled and wrapped shrubs and trees
- irrigation, runoff, drainage and overflow
- wind

Many weedy species grow in close association with important crop plants, such as cotton, corn (*Zea mays*), sorghum, and sugarcane (*Saccharum officinarum*). Therefore, some of the pathogens and insect pests of crop plants can be found on weedy species. The most visible example may be johnsongrass (*Sorghum halepense*), which can harbor a number of viruses and fungal pathogens that are also important on corn and sorghum. Since johnsongrass and large cereal grains grow in close proximity, the ease of movement for pest and pathogen (and their insect vectors) between the weed and the crops in their agricultural setting is logical and has been observed for many years.

Throughout this book we have attempted to identify some of the pathogens and pests that may be encountered on weedy species. These relationships are outlined beneath the plant descriptions. Also, several of the appendices attempt to identify some of the pathogens and insect pests that have been found on such weeds as sunflower (*Helianthus annuus*), giant reed (*Arundo donax*), bermudagrass, annual bluegrass (*Poa annua*), and johnsongrass. Where possible, the pathogens and pests found on the weed and on an agriculturally important crop are delineated. In many cases, however, a relationship has not been scientifically established and thus should not be extrapolated until such literature becomes available. Please consult your local county extension agent for detailed information concerning pests and pathogens that are important in your area.

Growth Habits of Weeds

In their growth habits weeds are categorized as annuals, biennials, or perennials. Annuals germinate from seed, grow vegetatively, flower, produce fruits and seeds, and die down entirely in one year or usually less. Winter annuals have seeds that germinate in the cooler fall months and grow vegetatively into the winter season. They flower, fruit, and produce seeds during the spring.

A biennial has a life cycle of two years, growing vegetatively during the first year and flowering and fruiting during the second year. Annuals and biennials are dependent upon seeds alone for reproduction and usually produce large numbers of seeds. We have not documented the occurrence of biennial weeds in the LRGV.

Perennial weeds live more than two years and typically cause the most pernicious problems. They have specialized stems called rhizomes, stolons, or bulbs that allow them to reproduce vegetatively. These stem modifications are highly resistant to environmental pressures, including drought and fire, and genetic variability is assured when the plants reproduce sexually.

Class Polypodiopsida

FERNS

SALVINIACEAE

Giant Salvinia, Annoyance Water Spangles

Salvinia molesta
D. S. Mitchell

Growth Habit: Perennial; a floating aquatic fern that lacks true roots.

Stems: Horizontal, below the water surface, simple or branched in mature stands.

Leaves: Simple, whorled with one submerged and rootlike and two floating, ovate or obovate, upper epidermis with distinctive "egg-beater" hairs (use 10× magnification).

Reproduction: This species is a sterile hybrid that reproduces vegetatively from stem fragments. Some submerged, filamentous leaves may produce sterile, ovoid sporocarps.

General Comments: Introduced. Giant salvinia is native to Brazil but has spread to many other freshwater wetlands of the world, including the United States (Barrett 1989). Giant salvinia is a federally listed noxious species that presently occurs in the northeastern portions of South Texas. It forms dense mats that completely cover the water surface to depths up to 0.5 m. The thick mats limit boating, increase stagnation, reduce water quality, and displace native plants and animals (DiTomaso and Healy 2003).

Biocontrol: The salvinia weevil (*Crytobagus salviniae*) has been used to control infestations in Africa, Asia, and Australia (DiTomaso and Healy 2003). Researchers at the USDA-APHIS laboratory in Edinburg, Texas, have been using the salvinia weevil to control giant salvinia infestations in southeastern Texas (Flores, Everitt, and Carlson 2006).

Class Magnoliopsida

DICOTS

AIZOACEAE

Winged Sea Purslane

Sesuvium verrucosum
C. Rafinesque-Schmaltz

Syn. *S. erectum*
D. Correll, *S. sessile*
B. Robinson,
non Persoon

Growth Habit: Annual; from a taproot.

Stems: Prostrate, glabrous, light green.

Leaves: Simple, opposite; blades linear, glabrous but with small papillae, succulent, margins entire; petioles with translucent margins near the base.

Inflorescence: Flowers solitary in the leaf axils.

> **Calyx:** Sepals 5, united below, green and papillose on the outside, reddish-purple and petaloid on the inner surface.

> **Corolla:** Petals absent.

> **Stamens:** 40–50, attached to a short disk around the ovary; filaments reddish-purple.

> **Pistil:** Ovary superior; styles 3.

Fruit: A many-seeded capsule, seeds black.

General Comments: Native. This species grows in dry saline or alkaline sites and where salinization is occurring.

Desert Horse Purslane, Verdolaga Blanca

Trianthema portulacastrum
C. Linnaeus

Growth Habit: Annual; from a taproot.

Stems: Prostrate, branching freely, mat forming, minutely pubescent.

Leaves: Simple, opposite; blades unequal, glabrous, succulent, broadly ovate or round, margins entire, reddish-purple; petioles and stipules present.

Inflorescence: Flowers solitary, axillary and sessile.

> **Calyx:** Sepals 5, free, but attached to the rim of the hypanthium, light pink.
> **Corolla:** Petals absent.
> **Stamens:** 8–10, attached to the rim of the hypanthium, light pink.
> **Pistil:** Ovary superior (perigynous); style 1, unbranched.

Fruit: A circumscissile, many-seeded capsule; seeds black, elliptic and mottled.

General Comments: Native. This species is common in fallow fields in late summer and fall. The U.S. Food and Drug Administration lists it as a toxic species. It is somewhat salt tolerant.

Biocontrol: Larvae of *Spoladea recurvalis* have been effective in controlling this species where broad-leaved crops are cultivated (Martin et al. 2001).

AMARANTHACEAE

Mat Chaff-flower

Alternanthera caracasana K. Kunth

Growth Habit: Annual or perennial; from a taproot.

Stems: Prostrate, mat forming, often turning red; pubescent with both branched and unbranched hairs.

Leaves: Simple, opposite; blades broadly elliptic to obovate, equal to unequal, glabrous, margins entire; petioles pubescent.

Inflorescence: Flowers in dense sessile, spiny, axillary clusters.

> **Calyx:** Bracts subtending the calyx spiny, light brown; sepals 5, tawny, with a mucronate apex.

> **Corolla:** Petals absent.

> **Stamens:** 5, anthers minute, yellow.

> **Pistil:** Ovary superior; style 1.

Fruit: A utricle with a single, round seed.

General Comments: Native. This species is common in weedy lawns, play-grounds, and golf courses during the warm season. Stiff spines are painful to the touch. The spiny propagules adhere to clothing, tires, and rubber-soled shoes.

Alligatorweed

Alternanthera philoxeroides
(K. von Martius)
A. Grisebach

Growth Habit: Perennial; aquatic, forming dense mats.

Stems: Branched or unbranched, hollow at the internodes, stoloniferous, rooting at the nodes, glabrous.

Leaves: Simple, opposite; blades linear, lanceolate, glabrous, margins entire; petioles absent.

Inflorescence: Flowers in a head or spike supported by an elongated peduncle.

> **Calyx:** Sepals 4, white, glabrous.
>
> **Corolla:** Absent.
>
> **Stamens:** 5.
>
> **Pistil:** Ovary superior; style elongate.

Fruit: A utricle; seeds inviable.

General Comments: Introduced. Alligatorweed has been found in a number of locations in South Texas, including drainage ditches and canals in the LRGV and in Lake Corpus Christi near Mathis. Dense floating mats impede water flow and boating in southeastern Texas and Louisiana. In the Houston area alligatorweed is often present as a garden weed and is difficult to eradicate.

Biocontrol: The exotic alligatorweed flea beetle (*Agasicles hygrophila*) has been used to control alligatorweed in a number of areas of Texas (Stutzenbaker 1999; Anonymous 2001).

Spreading Pigweed

Amaranthus blitoides
S. Watson

Growth Habit: Annual; from a taproot.

Stems: Prostrate, decumbent, minutely pubescent, often light pink.

Leaves: Simple, alternate; blades oblanceolate, glabrous, often with a white patch in the midsection, glaucous on the lower epidermis, margins often white, entire; petioles present.

Inflorescence: Flowers in axillary clusters, sessile.

Staminate Flowers:

Calyx: Bracts subtending calyx about twice the length of the sepals; sepals 5, free above but united near the base, with a dark green midsection, apex obtuse.

Corolla: Petals absent.

Stamens: 5, anthers yellow.

Pistillate Flowers:

Calyx: Similar to that of staminate flowers.

Corolla: Petals absent.

Pistil: Ovary superior; style branches usually 3.

Fruit: A utricle with a single, shiny, amber-colored seed.

General Comments: Introduced. This species is abundant throughout the growing season in dry sites, including lawns, fallow fields, and flower beds. Sperry et al. (1968) and Hart et al. (2003) indicate that all species of *Amaranthus* in Texas are toxic to livestock when the plants are mature but are palatable in the seedling stage.

Palmer Amaranth, Carelessweed, Quelite, Pigweed, Palmer's Pigweed

Amaranthus palmeri
S. Watson

Growth Habit: Annual; from a taproot.

Stems: Erect, usually turning red in older plants, glabrous.

Leaves: Simple, alternate; blades ovate to lanceolate, glabrous, margins entire; petioles grooved.

Inflorescence: Plants dioecious; male and female plants similar in appearance. Flowers in terminal spicate racemes and in axillary clusters below.

Staminate Flowers:

Calyx: Bracts subtending calyx green with acuminate apices; sepals 5, free with a green midvein and scarious margins.

Corolla: Petals absent.

Stamens: 5, anthers greenish-yellow.

Pistillate Flowers:

Calyx: Similar to that of staminate flowers.

Corolla: Petals absent.

Pistil: Ovary superior; style branches 2 or 3.

Fruit: A utricle with 1 shiny seed.

General Comments: Native. Carelessweeds and sunflowers are the most common dicot weeds in agricultural fields in South Texas. Each female plant produces millions of seeds that remain viable in the soil for years. High densities of this species interfere with the cotton harvest. The male plants are a major source of windblown pollen that causes respiratory allergies. See comments about toxicity under *A. blitoides*.

Tall Pigweed, Waterhemp, Tall Waterhemp

Amaranthus rudis
J. Sauer

Syn. *A. tamariscinus*
auct. (T. Nuttall)
A. Wood

Growth Habit: Annual; from a taproot; plants dioecious.

Stems: Erect, slightly ribbed, nearly glabrous, reddish tinged.

Leaves: Simple, alternate; blades ovate, glabrous, margins entire; petioles elongated.

Inflorescence: Flowers in terminal, spicate panicles and with some flowers in axillary clusters below.

> **Calyx:** Sepals 3, greenish, glabrous.
>
> **Corolla:** Petals absent.
>
> **Stamens:** 5.
>
> **Pistil:** Ovary superior; style branches 3.

Fruit: A utricle with a lustrous, round, black seed.

General Comments: Native. Tall pigweed is found occasionally in cracks in sidewalks and in crevices in asphalt pavement. See comments about toxicity under *A. blitoides*.

Drummond's Snakecotton

Froelichia drummondii
C. Moquin-Tandon

Growth Habit: Perennial or annual.

Stems: Erect, densely pubescent, leafless above.

Leaves: Simple, opposite; blades lanceolate, subsessile above and with elongated petioles below, pubescent, grayish-green, margins entire.

Inflorescence: Flowers in a mass of cottony hairs clustered near the stem apex.

> **Calyx:** Bracts subtending the sepals 2, papery, brown and included in cottony hairs; sepals 5, lobes linear.
>
> **Corolla:** Absent.
>
> **Stamens:** 5, anthers yellow.
>
> **Pistil:** Ovary superior; style 1.

Fruit: A 1-seeded utricle.

General Comments: Native. This species is common in deep, sandy soils on roadsides and in sandy, weedy pastures.

Woolly Tidestromia, Espanta Vaqueros

Tidestromia lanuginosa
(T. Nuttall) P. Standley

Growth Habit: Annual; from a taproot.

Stems: Erect, freely branching, becoming a tumbleweed at maturity, densely pubescent with branched hairs.

Leaves: Simple, opposite, unequal; blades round or broadly ovate, densely stellate pubescent, margins entire; petioles densely pubescent.

Inflorescence: Flowers in small, axillary clusters.

> **Calyx:** Sepals 5, greenish-translucent, free.
>
> **Corolla:** Petals absent.
>
> **Stamens:** 5, attached to a short hypanthium; anthers yellow.
>
> **Pistil:** Ovary semisuperior (perigynous); round.

Fruit: A 1-seeded utricle.

General Comments: Native. This heat-tolerant weed occurs on dry roadsides, on secondary dunes on Padre Island, and in a wide variety of disturbed sites. At maturity it becomes a tumbleweed.

APIACEAE

Wild Celery, Slimlobe Celery, Marsh Parsley

Cyclospermum leptophyllum
(C. Persoon)
T. A. Sprague *ex*
N. Britton &
Percy Wilson

Syn. *Apium leptophyllum*
(C. Persoon)
F. von Mueller *ex*
G. Bentham

Growth Habit: Annual; from a taproot.

Stems: Erect, but low growing, glabrous.

Leaves: Compound, alternate; blades dissected into parsleylike, linear lobes, glabrous; petioles elongated, translucent near the base.

Inflorescence: Flowers in dense axillary umbels.

> **Calyx:** Absent.
>
> **Corolla:** Petals 5, free, attached on the upper rim of the ovary, minute, white.
>
> **Stamens:** 5, anthers white.
>
> **Pistil:** Ovary inferior, ribbed, glabrous.

Fruit: A glabrous schizocarp.

General Comments: Introduced. Wild celery is a cosmopolitan, cool-season weed in agricultural fields, flower beds, and lawns.

Marsh Waterpennywort, Largeleaf Waterpennywort

Hydrocotyle bonariensis
P. Commerson *ex*
J. de Lamarck

Growth Habit: Perennial; forming colonies.

Stems: Low growing; rooting at the nodes, glabrous.

Leaves: Simple, alternate; blades round, peltate (with petiole attached to the midportion of the blade), glabrous, margins shallowly lobed.

Inflorescence: A compound umbel.

    Calyx: Sepals 5, inconspicuous.

    Corolla: Petals 5, free, yellowish-white.

    Stamens: 5, free.

    Pistil: Ovary inferior; styles 2.

Fruit: An elliptic schizocarp.

General Comments: Native. This species is a lawn weed in wet sites adjacent to faucet leaks. On Padre Island it is often abundant in wet depressions. In the Houston vicinity it is known as dollar weed.

ASCLEPIADACEAE

Primrose Milkweed, Longhorn Milkweed, Prairie Milkweed

Asclepias oenotheroides
A. von Chamisso &
D. von Schlechtendal

Growth Habit: Perennial; all parts with milky latex.

Stems: Erect, branching near the base, pubescent.

Leaves: Simple, opposite; blades ovate, pubescent, grayish-green on the lower epidermis, margins entire; petioles pubescent.

Inflorescence: An umbel from the upper leaf axils; pedicels pubescent.

> **Calyx:** Sepals 5, united below, pubescent on the outer surface, recurved.
>
> **Corolla:** Petals 5, partially united, recurved, glabrous, greenish.
>
> **Stamens:** Stamens 5 and attached to the stigmatic head; pollen in a waxy pollinium.
>
> **Pistil:** Ovary superior; styles 2.

Fruit: Ovary maturing into 2 follicles; seeds with soft, white hairs.

General Comments: Native. Many species of milkweeds are toxic, but it is not known if this species is poisonous. This species occurs in vacant lots and in a variety of disturbed sites. It attracts migrating butterflies, which serve as pollinating vectors.

ASTERACEAE

Field Ragweed, Weakleaf Bur Ragweed

Ambrosia confertiflora
A. P. de Candolle

Growth Habit: Perennial; forming large colonies from rhizomes; plants monoecious; malodorous.

Stems: Erect, pubescent.

Leaves: Simple, but dissected, alternate; blades lobed and secondarily lobed, appressed pubescent.

Inflorescence: Flowers unisexual; wind pollinated.

> **Staminate Florets:** In recurved heads; phyllaries united and pubescent; pollen yellow.

> **Pistillate Florets:** In small axillary clusters; involucre with scattered spines.

Fruit: A beaklike achene; pappus absent.

General Comments: Native. This species is more common in the drier areas of South Texas than other species of *Ambrosia*. The wind-borne pollen causes hay fever.

Pathogens and Pests: Several species of *Erysiphe* (*Golovinomyces*) *cichoracearum* (Romberg, Nujnez, and Farrar 2004), a powdery mildew, commonly infect cucurbits such as cantaloupe and watermelon in greenhouses and production fields. Ragweed located near agricultural fields may act as a reservoir host for this powdery mildew. However, in the LRGV, *Sphaerotheca fuliginea* is the most important powdery mildew fungus encountered on cucurbits (Marvin Miller, pers. comm.).

Western Ragweed, Perennial Ragweed, Smoothspike Ragweed

Ambrosia psilostachya
A. P. de Candolle

Growth Habit: Perennial; from rhizomes forming colonies; malodorous.

Stems: Erect, scabrous.

Leaves: Simple, alternate; blades lanceolate and dissected or lobed almost to the midvein, appressed pubescent.

Inflorescence: Flowers unisexual; wind pollinated.

> **Staminate Florets:** In terminal racemose heads, peduncles recurved, involucre of united phyllaries; pollen yellow.

> **Pistillate Florets:** Inconspicuous, in axillary clusters below the staminate florets.

Fruit: Achenes beaklike; pappus absent.

General Comments: Native. Western ragweed is the most common species of *Ambrosia* in South Texas. It represents the primary source of wind-borne pollen that causes hay fever.

Herbicidal Control: A broadleaf herbicide of 2,4-D and MCCP [2-(2-methyl-r-chlorophenoxy)] propionic acid controls young, actively growing plants. Combinations of dicamba and picloram are also options for controlling this weed.

Pathogens and Pests: See comments under *A. confertiflora*.

Giant Ragweed, Threelobed Ragweed

Ambrosia trifida
C. Linnaeus

Growth Habit: Annual; from a taproot; forming large stands; a resinous odor; produces copious amounts of wind-borne pollen.

Stems: Erect, slightly scabrous and ribbed, up to 3 m tall or taller.

Leaves: Simple, mostly opposite; blades deeply 3 to 5 lobed, mostly glabrous above and scabrous on the primary veins below, margins serrate; petioles elongate.

Inflorescence: Flowers unisexual; wind pollinated.

> **Staminate Florets:** In nodding, racemelike clusters above the inconspicuous pistillate florets; phyllaries united.

> **Pistillate Florets:** In small axillary clusters below the staminate florets; style branches 2.

Fruit: A beaklike achene; pappus absent.

General Comments: Native. Giant ragweed grows rapidly and forms dense stands in low sites. It is much more abundant in the northern sections of South Texas, where it represents a major source of wind-borne pollen that causes hay fever.

Pathogens and Pests: See comments under *A. confertiflora*.

Lowgrowing Lazydaisy, Poorland Lazydaisy, Lowgrowing Branched Lazydaisy

Aphanostephus ramosissimus
A. P. de Candolle

Growth Habit: Annual; from a taproot.

Stems: Decumbent from a basal rosette, pubescent.

Leaves: Basal leaves in a rosette, some broadly lobed; cauline leaves simple, alternate; blades lanceolate, narrowly lobed, pubescent; petioles absent.

Involucre: Phyllaries in several series, pubescent, purple at the apex.

Receptacle: Naked, convex; heads radiate.

> **Ray Florets:** Corollas white but purplish-white on the outer surface near the apex.
>
> **Disk Florets:** Corollas yellow.

Fruit: A minute cylindrical achene; pappus a minute fringed crown.

General Comments: Native. Species of *Aphanostephus* are common on deep sandy or sandy loam soils.

Fragrant Beggarticks

Bidens odorata
A. Cavanilles

Syn. *B. pilosa*
C. Linnaeus forma
odorata (A. Cavanilles)
E. Sherff

Growth Habit: Annual; with shallow roots.

Stems: Erect, square in cross section, ribbed and with scattered hairs.

Leaves: Pinnately compound with 3 to 5 leaflets, opposite; leaflets slightly pubescent, margins serrate-sinuate; petioles and rachis pubescent.

Involucre: Peduncle elongated, bearing several heads, pubescent; outer phyllaries recurved at the apex, pubescent.

Receptacle: Chaffy; heads radiate, fragrant.

> **Ray Florets:** Corollas white, slightly lobed at the apex, sterile, easily falling from the head.

> **Disk Florets:** Corollas yellow.

Fruit: An elongated, pubescent achene; pappus of 2 retrorsely barbed awns.

General Comments: Introduced. *Bidens odorata* is an attractive weed that has been introduced into South Texas with nursery stock, probably from commercial sources in Louisiana. We have not seen it in the South Texas landscape.

Pathogens and Pests: Dujovny, Usugi, and Shohara (1998) suggest that the mechanically and aphid-vectored (*Myzus* spp.) bidens mottle potyvirus (BMP) can be transferred to sunflower (*Helianthus annuus*). Artificial inoculation of *B. pilosa* with *Xylella fastidiosa* (the causal agent of citrus variegated chlorosis [CVC]) was 17 percent successful, suggesting that it is a possible weedy host for this phytoplasma (Lopes et al. 2002). CVC-causing *X. fastidiosa* is transmitted by the grass leafhopper (*Ferrariana trivittata*).

Prostrate Lawnflower, Hierba Del Caballo, Straggler Daisy

Calyptocarpus vialis
C. Lessing

Growth Habit: Perennial.

Stems: Sprawling, appressed pubescent.

Leaves: Simple, opposite; blades ovate to broadly elliptic, minutely pubescent; petioles pubescent.

Involucre: Phyllaries pubescent with appressed and longer hairs.

Receptacle: Heads radiate; chaff present.

> **Ray Florets:** Corollas yellow.

> **Disk Florets:** Corollas yellow; pappus similar to pappus of ray florets.

Fruit: An achene with a mottled surface; pappus of hornlike awns and occasionally with 1 shorter awn.

General Comments: Native. Prostrate lawnflower is one of the most common lawn weeds in South Texas. It competes with lawn grasses and thrives in shaded sites. Lawn mower blades should be adjusted to a high cutting position to allow turf grasses to shade-out this species.

Devilweed, Spiny Devilweed, Devilweed Aster

Chloracantha spinosa
(G. Bentham) G. Nesom

Syn. *Leucosyris spinosa*
(G. Bentham)
E. Greene; *Aster spinosus*
G. Bentham

Growth Habit: Perennial; from rhizomes, usually forming large colonies.

Stems: Erect, glabrous, slightly angled.

Leaves: If present, simple, alternate; blades awl shaped, glabrous.

Involucre: Phyllaries in 3 to 4 rows, glabrous, triangular.

Receptacle: Heads radiate; chaff absent.

> **Ray Florets:** Corollas white.
> **Disk Florets:** Corollas yellow.

Fruit: A minute achene; pappus of capillary bristles.

General Comments: Native. Devilweed forms large colonies in a variety of habitats but is more common in low, moist sites. It is a troublesome weed on rangelands in portions of southeastern Texas where soils have heavy clay content and high water-holding capacity.

Texas Thistle, Southern Thistle

Cirsium texanum
S. Buckley

Growth Habit: Annual; from a taproot.

Stems: Erect, ribbed with soft, appressed pubescence.

Leaves: Simple, alternate; blades broadly lanceolate, sessile and with a spiny auricle at the base, dark green on the upper epidermis with inconspicuous pubescence, lower epidermis grayish-green and densely appressed pubescent, margins deeply lobed and each lobe barbed with a sharp spine.

Involucre: Phyllaries in several series each with a white midvein and a reddish-purple, recurved, spine-tipped apex.

Receptacle: Chaff of soft, white bristles; heads discoid and large.

> **Ray Florets:** Absent.

> **Disk Florets:** Corollas pink or rose with 5 linear lobes.

Fruit: A glabrous, columnar achene; pappus of soft, white bristles.

General Comments: Native. Texas thistle is common on roadsides. The attractive heads are often illustrated in wildflower books.

Canadian Horseweed, Marestail

Conyza canadensis
(C. Linnaeus)
A. Cronquist

Growth Habit: Annual; from a taproot.

Stems: Erect, pubescent.

Leaves: Simple, alternate; blades linear, pubescent on the margins and midvein; petioles absent.

Involucre: Phyllaries linear, pubescent.

Receptacle: Naked; heads discoid.

> **Ray Florets:** Absent.

> **Disk Florets:** Corollas creamy white.

Fruit: A pubescent achene about 1 mm long; pappus of bristles about 2 mm long.

General Comments: Native. Canadian horseweed is common in agricultural fields, pastures, citrus orchards, fallow fields, and roadsides and is toxic to cattle and goats during drought conditions (Hart et al. 2003).

Pathogens and Pests: *Conyza canadensis* may serve as a host for *Sclerotinia minor,* a fungus causing sclerotinia blight of peanuts and other dicots in peanut-growing regions (Hollowell 2003).

Plains Coreopsis, Plains Tickseed, Dyer's Tickseed

Coreopsis tinctoria
T. Nuttall var. *tinctoria*

Growth Habit: Annual; from a taproot.

Stems: Erect, glabrous.

Leaves: Simple, in a rosette below and opposite with smaller lobes above; blades pinnately lobed with ovate lobes on lower leaves and linear lobes on upper leaves; petioles extended into a rachis with a few hairs at the base.

Involucre: Peduncles elongated; phyllaries in several series, the uppermost bronze above, apex obtuse.

Receptacle: Chaff present; heads radiate.

> **Ray Florets:** Corollas yellow with a red blotch near the base on the inner surface, sterile.

> **Disk Florets:** Corollas reddish-bronze.

Fruit: A winged, reddish-brown achene (the achenes resemble a scale insect or a tick); pappus of 2 minute awns.

General Comments: Native. Plains coreopsis is often abundant in damp, clay soils, including roadsides, field margins, and irrigation seepage areas. It is often illustrated in wildflower books.

Stiffleaf Scratchdaisy

Croptilon rigidifolium
(E. B. Smith) E. B. Smith

Syn. *C. divaricatum*
(T. Nuttall) C.
Rafinesque-Schmaltz
var. *hirtellum*
(L. Shinners) L. Shinners

Growth Habit: Annual; from a taproot.

Stems: Low growing, pubescent with some glandular hairs.

Leaves: Simple, alternate; blades oblanceolate, margins serrate and with stiff, spreading hairs.

Involucre: Phyllaries with scabrous margins and a conspicuous midvein.

Receptacle: Flat, chaff absent.

    Ray Florets: Corollas yellow, pistillate and fertile.

    Disk Florets: Corollas yellow, perfect and fertile.

Fruit: Achenes linear, dark; pappus of persistent bristles.

General Comments: Native. This is a low-growing species in deep, sandy soils, including secondary dunes on South Padre Island.

Yerba De Tago, Eclipta

Eclipta prostrata
(C. Linnaeus)
C. Linnaeus

Syn. *E. alba*
(C. Linnaeus) J. Hasskarl

Growth Habit: Annual.

Stems: Low growing and usually sprawling; appressed pubescent.

Leaves: Simple, opposite; blades lanceolate, appressed pubescent, margins remotely toothed; petioles appressed pubescent.

Involucre: Heads 1 to several in the leaf axils; pedicels densely pubescent; phyllaries pubescent.

Receptacle: Chaff present; heads radiate.

> **Ray Florets:** Corollas minute, white.

> **Disk Florets:** Corollas minute, white.

Fruit: A small, columnar achene; pappus of small, fringed bristles.

General Comments: Native. Yerba de tago is common in freshwater wetlands and is occasionally found as a weed in nurseries. It is more common in southeastern Texas, where it is often present in gardens.

Fosberg's Tasselflower

Emilia fosbergii
D. Nicholson

Growth Habit: Annual; from a shallow taproot.

Stems: Erect, densely pubescent below, sparsely pubescent above.

Leaves: Simple, alternate with auricle-like projections at the base; blades round, ovate, or broadly lanceolate near the base, reduced and linear-lanceolate above, pubescent primarily along the midvein, margins reddish-purple, remotely and bluntly toothed; petioles present on the lower lobes, pubescent but absent and blades clasping above.

Involucre: Phyllaries green, united nearly to the apex, minutely pubescent.

Receptacle: Naked; heads discoid.

> **Ray Florets:** Absent.

> **Disk Florets:** Corollas light reddish-orange.

Fruit: A cylindrical, pubescent achene; pappus of long, white, capillary bristles.

General Comments: Introduced. This exotic species has been introduced recently from Central or East Africa, and it has reached Texas from Mexico. It occurs in flower beds on the campus of the University of Texas–Pan American.

Spring Evax, Manystem Evax

Evax verna
C. Rafinesque-Schmaltz

Growth Habit: Annual; from a taproot.

Stems: Prostrate, branching freely with tomentose, matted pubescence.

Leaves: Simple, alternate; blades grayish-green, sessile with dense, matted pubescence.

Involucre: Phyllaries densely tomentose, heads included with a dense mass of white hairs.

Receptacle: Chaff present; heads discoid.

> **Ray Florets:** Absent.

> **Disk Florets:** Corollas inconspicuous.

Fruit: A minute achene; pappus absent.

General Comments: Native. Spring evax is a low-growing, cool-season weed on roadsides.

Threelobed Florestina

Florestina tripteris
A. P. de Candolle

Growth Habit: Annual; from a shallow taproot.

Stems: Erect, appressed pubescent.

Leaves: Trifoliolate, alternate; leaflets lanceolate to oblanceolate, pubescent; petioles pubescent.

Involucre: Phyllaries pubescent with white margins and a white, rounded apex.

Receptacle: Naked; heads discoid.

> **Ray Florets:** Absent.

> **Disk Florets:** Corollas white.

Fruit: A pubescent achene; pappus of erect scales.

General Comments: Native. This species is abundant in fallow fields throughout most of the growing season.

Pathogens and Pests: The leaflets are often covered with an unidentified species of a downy mildew.

Pennsylvania Cudweed

*Gamochaeta
pensilvanica*
(C. von Willdenow)
A. Cabrera

Syn. *Gnaphalium
pensilvanicum*
C. von Willdenow

Growth Habit: Annual; from a taproot.

Stems: Erect, tomentose-villous with white hairs.

Leaves: Simple, alternate; blades white tomentose, oblanceolate, light green above, silvery-grayish on the lower epidermis, margins entire.

Involucre: Phyllaries linear, greenish-translucent, subtended by a mass of white, woolly hairs.

Receptacle: Naked; heads discoid.

    Ray Florets: Absent.

    Disk Florets: Corollas inconspicuous.

Fruit: A grayish-brown achene; pappus of plumose bristles.

General Comments: Native. Pennsylvania cudweed is common in weedy lawns, flower beds, and fencerows during the cool season.

Broom Snakeweed, Broomweed, Ironweed

Gutierrezia sarothrae (F. Pursh) N. Britton & H. Rusby

Syn. *Xanthocephalum sarothrae* (F. Pursh) L. Shinners

Growth Habit: Perennial; woody at the base and shrublike.

Stems: Branched at the base and above.

Leaves: Simple, alternate; blades linear, glabrous and resinous.

Involucre: Phyllaries acute and usually resinous.

Receptacle: Flat or slightly convex.

> **Ray Florets:** Corollas yellow, 3 to 6 pistillate florets per head; pappus of minute scales.

> **Disk Florets:** Corollas yellow, 7 to 10 perfect, fertile florets per head; pappus longer than in ray florets.

Fruit: A densely pubescent achene.

General Comments: Native. Broom snakeweed is a rangeland weed in the north-central and western areas of South Texas. It is abundant in range sites that have been grazed heavily. It is toxic to livestock (Sperry et al. 1968; Hart et al. 2003).

Texas Snakeweed, Texas Broomweed

Gutierrezia texana
(A. P. de Candolle) J.
Torrey & A. Gray

Syn. *Xanthocephalum texanum*
(A. P. de Candolle)
L. Shinners

Growth Habit: Annual; from a taproot.

Stems: Erect with numerous lateral branches above, ribbed, with scattered papillae, viscid.

Leaves: Simple, alternate; blades linear, glutinous, margins entire, apex acute.

Involucre: Phyllaries lanceolate, glabrous, green.

Receptacle: Heads radiate; chaff present.

> **Ray Florets:** Corollas yellow.
> **Disk Florets:** Corollas yellow.

Fruit: A minutely pubescent achene; pappus scalelike, minute.

General Comments: Native. Texas snakeweed occurs in a wide variety of disturbed sites and is common in heavily compacted soils. It often forms dense stands in rangeland areas of South Texas.

Smallheaded Sneezeweed

Helenium microcephalum
A. P. de Candolle var.
oöclinium (A. Gray)
M. Bierner

Growth Habit: Annual; from a taproot.

Stems: Unbranched below and readily branching above.

Leaves: Simple, alter-
nate; blades elliptic to oblong-elliptic, resinous, margins serrate; petioles absent but leaf bases forming a lateral wing on the stems.

Involucre: Phyllaries in 2 series, resinous.

Receptacle: Columnar or globular.

> **Ray Florets:** Corollas yellow throughout.

> **Disk Florets:** Corollas reddish-brown.

Fruit: An achene with a pappus of scales.

General Comments: Native. This cool-season annual is often abundant in heavy clays on the margins of freshwater wetlands and other low sites. Sperry et al. (1968) and Hart et al. (2003) report that this species is highly toxic to cattle, horses, and sheep.

Sunflower, Annual
Sunflower, Mirasol

Helianthus annuus
C. Linnaeus

Growth Habit: Annual; from a taproot.

Stems: Erect with a resinous exudate when cut, scabrous-pubescent with stiff hairs and shorter, matted hairs, often blotched with purple coloration.

Leaves: Simple, alternate; blades ovate to broadly elliptic, acuminate at the apex, scabrous-pubescent only on the veins, margins entire; petioles grooved and scabrous.

Involucre: Peduncle elongated, scabrous; phyllaries in several series, acuminate-tapered at the apex, margins with stiff, perpendicular hairs.

Receptacle: Slightly convex; chaff purple, toothed at the apex.

    Ray Florets: Corollas yellow; florets pistillate but sterile.

    Disk Florets: Corollas yellow with purple teeth; florets fertile.

Fruit: A dark brown achene with appressed pubescence; pappus of 2 disarticulating awns.

General Comments: Native. This widely distributed species forms dense mono-typic stands during the crop-growing season as well as in fallow fields. It is often a co-dominant with *Amaranthus palmeri* (carelessweed). Sunflowers transpire high amounts of water required for agriculture, but they do not tolerate waterlogged soils.

Pathogens and Pests: See Appendix 2.

Silverleaf Sunflower

Helianthus argophyllus
J. Torrey & A. Gray

Growth Habit: Annual.

Stems: Erect, often 3 to 4 m tall, branched above, woolly-white.

Leaves: Simple, alternate; blades ovate, woolly-white, margins usually entire; petioles about as long as the blades.

Involucre: Peduncle elongated, woolly-white; phyllaries densely pubescent.

Receptacle: Nearly flat; chaff often purple and toothed at the apex.

> **Ray Florets:** Corollas yellow; florets pistillate but sterile.
>
> **Disk Florets:** Corollas brownish-yellow; florets fertile.

Fruit: Achenes pubescent; pappus of lanceolate awns.

General Comments: Native. Silverleaf sunflowers are tall, silvery, warm-season weeds in sandy pastures and are common in disturbed sites on North Padre Island.

Camphorweed

Heterotheca latifolia
S. Buckley

Growth Habit: Annual; from a taproot with a strong camphorlike odor.

Stems: Erect, pubescent.

Leaves: Simple, alternate; blades sessile, lanceolate-ovate, reduced above, glutinous, margins pubescent, entire.

Involucre: Peduncles with soft, ciliated hairs and shorter glandular hairs; phyllaries in several series, green with cilia and glandular hairs.

Receptacle: Chaff present; heads radiate.

> **Ray Florets:** Corollas yellow; pappus absent.

> **Disk Florets:** Corollas yellow; pappus with long, white bristles and a row of much shorter bristles.

Fruit: A pubescent achene.

General Comments: Native. Camphorweed is a warm-season species with a strong odor. It is abundant in fallow fields and is invasive in rangelands.

Poison Bitterweed, Western Bitterweed

Hymenoxys odorata
A. P. de Candolle

Growth Habit: Annual.

Stems: Erect, branching freely, with a few scattered hairs.

Leaves: Compound with several linear leaflets, alternate; leaflets grooved, slightly pubescent.

Involucre: Phyllaries pubescent.

Receptacle: Convex; florets subtended by a dense cluster of soft, white hairs; heads radiate.

   Ray Florets: Corollas yellow, pistillate, 3 lobed at the apex.

   Disk Florets: Corollas yellow, perfect.

Fruit: A silky pubescent achene; pappus of soft bristles.

General Comments: Native. Poison bitterweed occurs in the drier, western portions of South Texas. It is toxic to sheep and irritates the nose, eyes, and gastrointestinal tract of these animals (Sperry et al. 1968; Hart et al. 2003).

Common Jimmyweed, Common Goldenweed

Isocoma coronopifolia
(A. Gray) E. Greene

Growth Habit: Perennial; a low-growing, semiwoody shrub.

Stems: Erect, resinous, woody near the base and herbaceous above.

Leaves: Simple, alternate or in fascicles; blades linear, glabrous, margins usually pinnately lobed.

Involucre: Phyllaries resinous, in several rows.

Receptacle: Slightly elevated.

>**Ray Florets:** Absent.

>**Disk Florets:** Yellow.

Fruit: A pubescent achene; pappus of bristles.

General Comments: Native. Common jimmyweed is abundant in heavily grazed rangelands in the drier, western portions of South Texas, where it often creates a management problem.

Drummond's Jimmyweed, Drummond's Goldenweed

Isocoma drummondii
(J. Torrey & A. Gray)
E. Greene

Growth Habit: Perennial; a low-growing, semiwoody shrub.

Stems: Erect, resinous, woody near the base and herbaceous above.

Leaves: Simple, alternate; blades linear, glabrous, resinous, margins usually entire.

Involucre: Phyllaries resinous and in 3 or 4 rows.

> **Ray Florets:** Absent.

> **Disk Florets:** Yellow.

Fruit: A 4-angled pubescent achene; pappus of light brown bristles.

General Comments: Native. This species is common in heavily grazed coastal clay soils. It flowers from October until late December.

Seacoast Sumpweed, Marshelder, Pelocote

Iva annua C. Linnaeus

Growth Habit: Annual; from a taproot; aromatic.

Stems: Erect, scabrous with antrorsely oriented hairs.

Leaves: Simple, alternate above, opposite below; blades ovate to lanceolate with 3 distinct veins arising from the petioles, margins toothed, apex acuminate.

Inflorescence: Bracts subtending heads lanceolate with ciliate margins; heads in spikes; discoid.

> **Ray Florets:** Absent.

> **Disk Florets:** Corollas green, inconspicuous; each head with several males and 1–2 female florets; wind pollinated.

Fruit: An achene.

General Comments: Native. *Iva annua* is fairly common in heavy, clay soils on the margins of wetlands. The wind-borne pollen causes hay fever.

Prickly Lettuce

Lactuca serriola C.
Linnaeus

Growth Habit: Annual;
all parts with a milky
latex.

Stems: Erect, often
whitish above, glabrous.

Leaves: Simple, alternate;
blades linear-lanceolate,
pinnately lobed or with
entire, prickly margins,
auriculate and clasping at
the base, upper epidermis
glabrous, midvein of the lower epidermis prickly.

Involucre: Phyllaries in several rows.

Receptacle: Naked; heads ligulate.

> **Ray Florets:** Corollas yellow with a pinkish tint.
>
> **Disk Florets:** Absent.

Fruit: A glabrous, minutely winged achene; pappus of soft, silky hairs.

General Comments: Introduced. Prickly lettuce occurs in dry, disturbed sites and
fallow fields during late spring and early summer. The ray florets open on sunny
mornings and usually close by early afternoon.

Texas Palafoxia

Palafoxia texana
A. P. de Candolle var.
texana

Growth Habit: Annual; from a taproot.

Stems: Erect, appressed pubescent.

Leaves: Simple, alternate; blades lanceolate, pubescent with pustulate-based hairs, margins entire; petioles pubescent.

Involucre: Phyllaries in 1–2 rows, pubescent.

Receptacle: Naked; heads discoid.

>**Ray Florets:** Absent.
>
>**Disk Florets:** Corollas rose-lavender, tubular below and lobed above; anthers black.

Fruit: A pubescent achene; pappus of scales.

General Comments: Native. Texas palafoxia is abundant in fallow fields, where it is often associated with *Florestina tripteris*. *Palafoxia texana* is represented by var. *ambigua* (L. Shinners) B. L. Turner & M. Morris in the LRGV. It has a slightly shorter pappus than var. *texana*.

False Ragweed, Santa Maria Feverfew, Cicutilla

Parthenium hysterophorus
C. Linnaeus

Growth Habit: Annual; from a taproot.

Stems: Erect, ribbed, pubescent.

Leaves: Simple, alternate; blades deeply dissected and lobed, but linear with entire margins above, appressed pubescent; petioles grooved, pubescent.

Involucre: Peduncle elongated, pubescent, and bearing numerous heads; phyllaries pubescent.

Receptacle: Chaff translucent; heads radiate.

> **Ray Florets:** Corollas white, inconspicuous.

> **Disk Florets:** Corollas white; florets sterile.

Fruit: A glabrous, flattened achene; pappus with 2 awns.

General Comments: Native. False ragweed has leaves similar to those of species of *Ambrosia*. Pollen of a genotype in India causes allergic reactions (Towers 1979). It is listed as an invasive species in Australia and India.

Purple Marshfleabane, Canela

Pluchea purpurascens
(O. Swartz)
A. P. de Candolle

Syn. *P. odorata*
A. H. Cassini

Growth Habit: Annual; shallowly rooted.

Stems: Erect, nearly glabrous below, but pubescent near the apex.

Leaves: Malodorous; simple, alternate; blades ovate or ovate-lanceolate, glabrous, margins entire or serrated; petioles elongated.

Involucre: Phyllaries pubescent.

Receptacle: Flat.

   Ray Florets: Absent.

   Disk Florets: Corollas purple or rose colored.

Fruit: An angled achene; pappus of bristles.

General Comments: Native. Purple marshfleabane is common on the margins of freshwater and brackish wetlands.

Smallflowered False Dandelion, Texas Dandelion

Pyrrhopappus pauciflorus
(D. Don)
A. P. de Candolle

Growth Habit: Annual; from a taproot, all parts with a milky latex.

Stems: Erect, pubescent to nearly glabrous.

Leaves: Simple, alternate, pubescent; blades oblanceolate, margins deeply lobed or remotely toothed; petioles, if present, pubescent.

Involucre: Phyllaries in 2–3 rows.

Receptacle: Naked; heads ligulate.

> **Ray Florets:** Corollas yellow, pubescent in the throat.
>
> **Disk Florets:** Absent.

Fruit: A brown, beaked achene with longitudinal grooves and horizontal striations; pappus of plumose hairs.

General Comments: Native. This attractive weedy species is similar to *Taraxacum officinale* (dandelion). It is common in lawns and on roadsides during the cool season.

Camphor Tansyaster, Leafyhead Tansyaster

Rayjacksonia phyllocephala
(A. P. de Candolle)
R. L. Hartman &
M. L. Lane

Syn. *Machaeranthera phyllocephala*
(A. P. de Candolle)
L. Shinners

Growth Habit: Annual; from a taproot, with a strong camphorlike odor.

Stems: Erect, often reddish, pubescent.

Leaves: Simple, alternate, sessile; blades lanceolate or linear, pubescent and glandular, margins toothed with spine-tipped bristles.

Involucre: Phyllaries in several series, glandular.

Receptacle: Chaff present; heads radiate.

> **Ray Florets:** Corollas yellow.

> **Disk Florets:** Corollas yellow.

Fruit: A densely pubescent achene; pappus of numerous white hairs.

General Comments: Native. Camphor tansyaster occurs in alkaline and saline sites. It is abundant near the coast and is present on the margins of storm surge tidal flats on Padre Island. The camphorlike odor of the crushed leaves is similar to the odor of *Heterotheca latifolia*.

Texas Groundsel, Texas Squawweed

Senecio ampullaceus
W. J. Hooker

Growth Habit: Annual; from a shallow taproot.

Stems: Erect, glabrous.

Leaves: Simple, alternate; blades lanceolate or lanceolate-ovate, sessile, densely pubescent with tomentose hairs (similar to a spider web pattern), margins sinuate; petioles absent, blades clasping.

Involucre: Peduncle with reduced linear leaves; phyllaries green below and purple tinged at the apex.

Receptacle: Naked; heads radiate.

> **Ray Florets:** Corollas yellow, strap shaped, pistillate.

> **Disk Florets:** Corollas yellow.

Fruit: A columnar, pubescent achene; pappus of numerous silky hairs.

General Comments: Native. Texas groundsel is an endemic species common in sandy pastures and on roadsides during the cool season. Livestock may develop cirrhosis of the liver by consuming this species (Hart et al. 2003).

Pathogens and Pests: Some species of *Senecio* have been reported to be a host for the fungus *Albugo tragopogonis,* which causes "white rust" disease on a number of horticultural species (Farr et al. 1989).

Groundsel, Ragwort, Squawweed

Senecio tampicanis
A. P. de Candolle

Growth Habit: Annual; from a taproot.

Stems: Erect, glabrous.

Leaves: Simple and pinnately lobed, in a rosette below and alternate and reduced above, glabrous; petiole extended into a rachis.

Involucre: Peduncle with minute, leaflike enations; phyllaries green, acute at the apex.

    Ray Florets: Corollas yellow, strap shaped, pistillate.

    Disk Florets: Corollas yellow.

Fruit: A columnar, pubescent achene; pappus of numerous silky hairs.

General Comments: Native. Groundsel forms large colonies during the cool season along roadsides and in damp soils in Cameron County.

Pathogens and Pests: See comments under *S. ampullaceus.*

Annual Sowthistle

Sonchus oleraceus
C. Linnaeus

Growth Habit: Annual; from a taproot, all parts with a milky latex.

Stems: Erect, glabrous, often reddish-purple tinged.

Leaves: A rosette below, simple, alternate above, pinnatisect or pinnatifid, upper-most leaves with a rounded auricle and an acute projection near the base, margins spine tipped.

Involucre: Phyllaries in at least 3 series, glabrous, margins translucent.

Receptacle: Naked; heads ligulate.

   Ray Florets: Corollas yellow.

   Disk Florets: Absent.

Fruit: A brown, rough-surfaced achene; pappus of soft, fluffy hairs.

General Comments: Introduced. Annual sowthistle is present in a variety of disturbed sites during the cool season. It is the ecological equivalent of dandelions in South Texas.

Wireweed, Saltmarsh Aster

Symphyotrichum divaricatum (T. Nuttall) G. Nesom

Syn. *Aster subulatus* (T. Nuttall) J. Torrey & A. Gray

Growth Habit: Annual; from a taproot.

Stems: Erect, ranging from several centimeters tall in lawns to several meters tall in weedy sites; appressed pubescent primarily at the nodes.

Leaves: Simple, alternate; blades linear-lanceolate to linear on branches, glabrous or nearly so and with a prominent midvein, margins entire; petioles absent.

Involucre: Phyllaries in several series, recurving at maturity, apex purple tipped; heads radiate.

Receptacle: Flat or slightly convex.

> **Ray Florets:** Corollas pistillate, white or light purple.

> **Disk Florets:** Corollas yellow.

Fruit: An achene; pappus of capillary bristles.

General Comments: Native. This species is common in wetlands, fallow fields, and lawns.

Dandelion

Taraxacum officinale
G. Weber *ex* F. Wiggers

Growth Habit: Annual or perennial; from a thick taproot; all parts with a milky latex.

Stems: Low growing and branching.

Leaves: Simple, in a basal rosette; blades pinnatifid, glabrous, margins dissected and toothed.

Involucre: Scape elongated in fruit; phyllaries glabrous and in 2 series, the outer shorter and recurved and the inner erect and longer.

Receptacle: Slightly concave.

> **Ray Florets:** Corollas yellow, numerous.

> **Disk Florets:** Absent.

Fruit: An achene with a plumed, parachute-like pappus readily dispersed by wind.

General Comments: Introduced. Dandelions are rare in South Texas, but a small population has persisted for many years in McAllen (Hidalgo County). Dandelion is one of the most common species in temperate regions of the globe.

Tridax Daisy, Coat Buttons

Tridax procumbens
C. Linnaeus

Growth Habit: Annual; from an enlarged taproot.

Stems: Sprawling, pubescent, rooting at the nodes.

Leaves: Simple, opposite; blades ovate with pustulate-based hairs, margins toothed; petioles pubescent.

Involucre: Peduncle elongated, pubescent, bearing a solitary head; phyllaries pubescent, green below and reddish tinged above.

Receptacle: Chaff present, reddish tinged along the midvein above; heads radiate.

> **Ray Florets:** Corollas light yellow or cream colored, pistillate, 3 lobed with 1 short pointed lobe.

> **Disk Florets:** Corollas yellow above and reddish tinged and yellow below.

Fruit: A pubescent achene; pappus of soft, white bristles.

General Comments: Introduced. Tridax daisy, a federally listed noxious weed, was reported in Weslaco, Texas (Brown and Elsik 2002), and we have seen it in a recently landscaped park in Edinburg, Texas.

Cowpen Daisy, Golden Crownbeard, Butter Daisy

Verbesina encelioides
(A. Cavanilles)
G. Bentham & J. Hooker
ex A. Gray

Growth Habit: Annual; from a taproot.

Stems: Erect, appressed pubescent.

Leaves: Simple, alternate; blades broadly lanceolate to broadly elliptic, grayish-green, pubescent, margins dentate-serrate, base truncate or occasionally cordate; petioles appressed pubescent.

Involucre: Phyllaries numerous, linear, pubescent, 2/3 the length of the rays.

Receptacle: Slightly convex, chaff yellow near the apex; heads radiate.

> **Ray Florets:** Corolla yellow, 3 lobed at the apex, tube and throat pubescent, pistillate.
>
> **Disk Florets:** Corolla yellow, tube pubescent below.

Fruit: A flattened, pubescent achene; pappus of short hairs and 2 soft bristles.

General Comments: Native. Cowpen daisy is common in cultivated fields. It is toxic to livestock (Kingsbury 1964). Nectar from this weed ruins the quality of honey in commercial beehives.

Common Cocklebur, Abrojo

Xanthium strumarium
C. Linnaeus

Growth Habit: Annual; from a taproot; plants monoecious.

Stems: Erect, hispid.

Leaves: Simple, alternate or opposite near the base; blades hispid, margins toothed or lobed; petioles elongate.

Inflorescence: Heads unisexual; pistillate heads in the midaxis; staminate heads in the upper axis.

> **Staminate Florets:** Smaller than the pistillate florets; ray florets absent; disk florets cream colored.

> **Pistillate Florets:** Burlike; involucre chaffy; corolla and pappus absent.

Fruit: An achene enclosed within an enlarged bur.

General Comments: Native. Cockleburs occur occasionally in damp sites during the warm season and late fall, but this weedy species is more common in the temperate zone. The burs adhere readily to clothing and passing animals. Cockleburs are toxic in the seedling stage, but the mature burs are not consumed by livestock (Sperry et al. 1968). However, cattle may be poisoned by hay contaminated with cocklebur seeds (Hart et al. 2003). Two achenes are usually present in each bur.

Palmer's Goldenweed, False Broomweed

Xylothamia palmeri
(A. Gray) G. Nesom

Syn. *Ericameria austrotexana*
M. C. Johnston

Growth Habit: Perennial; from a taproot; a low shrub.

Stems: Densely branched, glabrous, resinous.

Leaves: Simple, alternate; blades linear, glabrous, resinous, margins entire.

Involucre: Phyllaries resinous with a dark spot near the apex.

Receptacle: Flat or slightly elevated.

> **Ray Florets:** Often absent; pale yellow and inconspicuous if present.
>
> **Disk Florets:** Pale, yellowish-white; pappus of bristles.

Fruit: A hirsute achene.

General Comments: Native. This species is a noxious weed on well-drained clay and clay loam soils on rangelands (Mayeux, Scifres, and Crane 1980).

Asiatic Hawkweed

Youngia japonica
(C. Linnaeus)
A. P. de Candolle

Growth Habit: Annual; from a taproot.

Stems: Absent or obscure; all parts with a milky latex.

Leaves: Simple, in a clustered rosette; blades lyrate or pinnatifid, broadest near the apex, nearly glabrous above but pubescent on the veins of the lower epidermis, margins on lower leaves lobed to the midvein; petioles elongated, pubescent.

Involucre: Inflorescence supported by an elongated peduncle bearing a cluster of heads and arising from the plant base; phyllaries in 2 series, nearly translucent on the margins.

Receptacle: Naked.

> **Ray Florets:** Corollas yellow with 5 teeth at the apex; pappus of soft, white hairs; ovary with a cartilaginous band at the apex.

> **Disk Florets:** Absent.

Fruit: A ribbed achene.

General Comments: Introduced. This species from eastern Asia is a weed in flower beds. It was probably introduced with nursery seedlings in potting soil. It is a common lawn weed in the Houston area.

BORAGINACEAE

Curacao Heliotrope, Alkali Heliotrope, Seaside Heliotrope, Salt Heliotrope

Heliotropium curassavicum
C. Linnaeus

Growth Habit: Perennial.

Stems: Prostrate, glabrous, grayish-green.

Leaves: Simple, alternate or occasionally opposite; blades linear-lanceolate, glabrous, succulent, margins entire; petioles inconspicuous.

Inflorescence: An axillary or terminal scorpioid cyme.

> **Calyx:** Sepals 5, united below, glabrous, grayish-green.
>
> **Corolla:** Petals 5, united, white above and greenish-white in the tube.
>
> **Stamens:** 5, epipetalous; filaments absent.
>
> **Pistil:** Ovary superior; stigma cap shaped.

Fruit: A 4-lobed schizocarp separating into 4 bony nutlets.

General Comments: Native. This salt-tolerant species and *Sporobolus pyramidatus* (whorled dropseed) are indicators of salinization They are usually the only species present on the margins of salty areas in agricultural fields.

Indian Heliotrope

Heliotropium indicum
C. Linnaeus

Growth Habit: Annual; from a taproot.

Stems: Erect, branched above the base, pubescent.

Leaves: Simple, alternate; blades ovate or elliptic, pubescent; margins undulate.

Inflorescence: A scorpioid cyme with 2 rows of densely crowded flowers and fruits.

>**Calyx:** Sepals 5, unequal, united below.
>
>**Corolla:** Petals 5, pubescent, united, blue-violet or occasionally white.
>
>**Stamens:** 5, attached to the corolla.
>
>**Pistil:** Ovary superior, divided into 4 lobes; style unbranched.

Fruit: 4 bony nutlets.

General Comments: Native. Indian heliotrope is often present on the margins of wetlands.

BRASSICACEAE

Chinese Mustard

Brassica juncea
(C. Linnaeus)
V. Czernajew

Growth Habit: Annual; from a taproot.

Stems: Erect, branching, up to 1.5 m tall.

Leaves: Simple, alternate; blades lyrate-pinnatifid below and linear above; petioles present on all leaves.

Inflorescence: An elongated raceme; pedicels 10–12 mm long.

> **Calyx:** Sepals 4, free, erect at maturity, green.

> **Corolla:** Petals 4, yellow, showy with conspicuous veins.

> **Stamens:** 6, in a 4 long and 2 short arrangement; anthers yellow.

> **Pistil:** Ovary superior; style 1.

Fruit: An elongated, slightly pubescent, 2-chambered capsule; style persistent.

General Comments: Introduced. Chinese mustard is a common roadside weed in the northern portions of South Texas during the winter months.

Tansymustard

Descurainia pinnata
(T. Walter) N. Britton

Growth Habit: Annual; from an elongated taproot.

Stems: Erect, pubescent with glandular and nonglandular hairs.

Leaves: Bipinnately compound, alternate, reduced above, stellate pubescent.

Inflorescence: An elongated, terminal raceme; pedicels glabrous.

    Calyx: Sepals 4, free, yellowish.

    Corolla: Petals 4, free, yellow.

    Stamens: 6, not all the same length.

    Pistil: Ovary superior; style 1.

Fruit: An elongated capsule.

General Comments: Native. Tansymustard is a cool-season weed on sandy roadsides and in sandy fields and pastures. Hart et al. (2003) report that this species may cause nitrate poisoning in livestock.

Wright's Woollyfruit Pepperwort

Lepidium lasiocarpum
T. Nuttall *ex* J. Torrey &
A. Gray var. *wrightii*
(A. Gray) C. Hitchcock

Growth Habit: Annual; from a taproot.

Stems: Erect, several arising from the base, pubescent with hairs of 2 sizes.

Leaves: Simple, alternate; blades lobed below and serrate above, oblanceolate, pubescent.

Inflorescence: A raceme; pedicels pubescent.

    Calyx: Sepals 4, free.

    Corolla: Petals 4, white, minute or absent.

    Stamens: 6, in a 4 + 2 arrangement, anthers white.

    Pistil: Ovary superior with a small notch at the apex.

Fruit: A 2-seeded, round capsule.

General Comments: Native. This species is a cool-season weed in lawns and in a variety of disturbed sites.

Pathogens and Pests: Species of *Lepidium* are hosts for the root-decaying fungus *Leptosphaeria maculans* (*Phoma lingam*), which causes "black-leg" disease of cruciferous vegetables (Balesdent et al. 1998).

Woollypod Bladderpod

Lesquerella lasiocarpa
(W. J. Hooker *ex*
A. Gray) S. Watson var.
lasiocarpa

Growth Habit: Annual; from a taproot.

Stems: Prostrate, pubescent with smaller stellate hairs and longer, usually unbranched hairs.

Leaves: Simple, alternate; blades lanceolate, stellate pubescent, margins toothed; petioles absent.

Inflorescence: Flowers in a widely spaced raceme; pedicels, in fruit, curved downward.

>**Calyx:** Sepals 4, free, stellate pubescent.

>**Corolla:** Petals 4, free, bright yellow with folded appendages at the base.

>**Stamens:** 6, in a 4 + 2 arrangement.

>**Pistil:** Ovary superior, densely pubescent; style 1.

Fruit: A round, but nearly flattened capsule with a visible longitudinal septum; seeds disk shaped.

General Comments: Native. *Lesquerella lasiocarpa* forms dense carpets of attractive yellow flowers on roadsides and in sandy loam fields during late winter and early spring.

Large Selenia, Rio Grande Valley Selenia

Selenia grandis R. Martin

Growth Habit: Annual; from a taproot.

Stems: Erect or nearly prostrate in older plants, glabrous.

Leaves: Simple, in a basal rosette below, alternate above; blades intricately lobed, glabrous; petiole extended into a rachis.

Inflorescence: Flowers in a raceme with elongated pedicels.

> **Calyx:** Sepals 5, free, light green on the outside and light yellow within.
>
> **Corolla:** Petals 5, free, golden, fragrant.
>
> **Stamens:** 6, 4 long and 2 slightly shorter; filaments yellow; anthers greenish and elliptical when pollen is released.
>
> **Pistil:** Ovary superior with green, lateral appendages at the base; style 1, unbranched.

Fruit: An oblong, beaked capsule with irregularly arranged purple blotches; seeds winged, numerous.

General Comments: Native. This attractive, fragrant species is a cool-season weed in carrot, cabbage, and onion fields south of Alamo, Texas (Hidalgo County).

Virginia Winged Rockcress

Sibara virginica
(C. Linnaeus) R. Rollins

Growth Habit: Annual; from a taproot.

Stems: Erect, pubescent below but nearly glabrous above.

Leaves: Simple, in a basal rosette below, alternate above; blades pinnatifid, glabrous.

Inflorescence: A terminal raceme.

> **Calyx:** Sepals 4, free, with a few scattered hairs.
>
> **Corolla:** Petals 4, free, white.
>
> **Stamens:** 6, in a 4 + 2 arrangement; anthers yellow.
>
> **Pistil:** Ovary superior.

Fruit: A linear capsule with numerous seeds.

General Comments: Native. Virginia winged rockcress is a weed in lawns, flower beds, and nurseries during the cool season.

London Rocket

Sisymbrium irio
C. Linnaeus

Growth Habit: Annual; from a taproot.

Stems: Erect, glabrous.

Leaves: Simple, alternate; blades dissected to the midvein, nearly glabrous but with minute hairs on the veins above; petioles with scattered hairs.

Inflorescence: A raceme.

    Calyx: Sepals 4, free, green.

    Corolla: Petals 4, free, light yellow.

    Stamens: 6, with 4 long and 2 short filaments.

    Pistil: Ovary superior; style 1.

Fruit: An elongated capsule with numerous seeds.

General Comments: Introduced. London rocket is an abundant cool-season weed in agricultural fields, on roadsides, and in other disturbed sites.

CAMPANULACEAE

Berlandier's Lobelia

Lobelia berlandieri
A. L. de Candolle

Growth Habit: Annual; from a shallow taproot.

Stems: Erect, glabrous above and minutely pubescent below and minutely winged.

Leaves: Simple, alternate; blades ovate, minutely pubescent, margins toothed with white glands on the teeth; petioles pubescent on the margins.

Inflorescence: Flowers in a raceme; pedicels pubescent.

>**Calyx:** Sepals 5, united below, linear, glabrous.

>**Corolla:** Petals 5, united, zygomorphic, bluish-purple and white with glandular pubescence near the base within the tube.

>**Stamens:** 5, epipetalous; filaments free and anthers united around the style; staminodes 2, green.

>**Pistil:** Ovary perigynous, style 1, stigma with brushlike pubescence.

Fruit: A many-seeded capsule.

General Comments: Native. *Lobelia berlandieri* contains alkaloids that are toxic to cattle and goats (Kingsbury 1964; Hart et al. 2003). In the spring, following wet winters, it forms carpets of brilliant blue flowers in the coastal mainland areas of South Texas.

CAPPARIDACEAE

Rio Grande Clammyweed

Polanisia dodecandra
(C. Linnaeus)
A. P. de Candolle subsp.
riograndensis H. Iltis

Growth Habit: Annual; from a taproot; plants malodorous.

Stems: Erect, pubescent with glandular-tipped hairs.

Leaves: Trifoliolate, alternate; leaflets ovate-lanceolate, pubescent, margins entire; petioles with hairs similar to hairs on stems.

Inflorescence: A raceme with pedicels subtended by green bracts.

> **Calyx:** Sepals 4, free, 3 about the same length and 1 larger, pinkish, pubescent.

> **Corolla:** Petals 4, free, clawed, pinkish-rose; slightly zygomorphic.

> **Stamens:** Numerous (more than 10); filaments pink.

> **Pistil:** Ovary superior; style 1; a truncate gland present adjacent to the ovary.

Fruit: An elongated, pubescent capsule; seeds round.

General Comments: Native. Rio Grande clammyweed is an attractive but ill-smelling weed common in the drier, western portions of South Texas.

CARYOPHYLLACEAE

Chickweed

Stellaria media
(C. Linnaeus) D. Villars

Growth Habit: Annual; from a taproot.

Stems: Low growing, creeping or sprawling, with a longitudinal row of hairs.

Leaves: Simple, opposite; blades ovate, glabrous, margins entire; petioles grooved with a few scattered hairs.

Inflorescence: 1 to several flowers from the leaf axils.

> **Calyx:** Sepals 5, free, pubescent, margins translucent.

> **Corolla:** Petals 5, bifurcate nearly to the base, free, white, shorter than the sepals.

> **Stamens:** 7–10; anthers purplish-white.

> **Pistil:** Ovary superior; styles 3.

Fruit: A capsule; seeds round with a mottled surface.

General Comments: Introduced. Chickweed is a rare temperate zone weed in South Texas. It has been introduced with turf grasses in shaded locations on the University of Texas–Pan American campus.

Pathogens and Pests: Tomato spotted wilt virus (TSWV) is transmitted by tobacco thrips (*Frankliniella fusca*) from *Stellaria* to other weedy species and crop plants. *Stellaria media* is an "over season" host for the thrips (Groves et al. 2001, 2002). Also, chickweeds may serve as a host for *Sclerotinia minor,* a fungus causing sclerotinia blight of peanuts and other dicots in peanut-growing areas (Hollowell 2003).

CHENOPODIACEAE

Crested Saltbush, Quelite Saltbush, Sand Atriplex, Seabeach Orach

Atriplex pentandra
(N. von Jacquin)
P. Standley

Syn. *A. arenaria*
T. Nuttall

Growth Habit: Annual or perennial; plants monoecious.

Stems: Erect, leaning, slightly ribbed, glabrous.

Leaves: Simple, alternate; blades lanceolate, glabrous, green above and glaucous-scurfy on the lower epidermis, margins undulate and toothed; petioles present.

Inflorescence: Staminate flowers clustered on an elongated axis; pistillate flowers solitary or few in the leaf axils.

> **Staminate Flowers:**
>
> > **Calyx:** Sepals 5, each with a green midvein.
> >
> > **Corolla:** Petals absent.
> >
> > **Stamens:** 5, pollen creamy-yellow.
>
> **Pistillate Flowers:**
>
> > **Calyx:** Indistinct with united, spiny bracts.
> >
> > **Corolla:** Petals absent.
> >
> > **Pistil:** Ovary superior, enclosed by spiny bracts.

Fruit: A utricle with 1 seed.

General Comments: Native. Crested saltbush occurs on the margins of salt marshes primarily in the coastal counties of South Texas.

Berlandier's Goosefoot

Chenopodium berlandieri
C. Moquin-Tandon

Growth Habit: Annual; from a taproot; malodorous.

Stems: Erect, glabrous, often reddish-purple at the nodes.

Leaves: Simple, alternate; blades ovate and broadest near the base, scurfy especially on the lower epidermis, margins wavy or irregular sinuate; petioles elongated.

Inflorescence: Flowers in densely clustered terminal panicles.

    Calyx: Sepals 5, green, mealy or scurfy.

    Corolla: Petals absent.

    Stamens: 5; anthers cream colored.

    Pistil: Ovary superior.

Fruit: A utricle.

General Comments: Native. Berlandier's goosefoot is a foul-smelling, cool-season weed that occurs in a wide variety of disturbed sites. The wind-borne pollen may cause respiratory allergic reactions.

Nettleleaf Goosefoot

Chenopodium murale
C. Linnaeus

Growth Habit: Annual; from a taproot; odorless.

Stems: Erect, glabrous, often reddish tinged.

Leaves: Simple, alternate; blades ovate, glabrous or mealy, margins broadly toothed; petioles elongated, grooved.

Inflorescence: Flowers in axillary and terminal panicles.

> **Calyx:** Sepals 5, green, mealy with blunt-pointed projections at the apex.

> **Corolla:** Petals absent.

> **Stamens:** 5; anthers white.

> **Pistil:** Ovary superior.

Fruit: A black or amber-colored utricle.

General Comments: Introduced. This species is similar to *C. berlandieri,* but the foliage is odorless. It occurs in a wide variety of disturbed sites.

Prickly Russian Thistle, Tumbleweed

Salsola tragus
C. Linnaeus

Syn. *S. iberica* F. Sennen
& C. Pau

Growth Habit: Annual; from a taproot.

Stems: Erect, glabrous with alternate green and white lines, turning red and finally brown and becoming a tumbleweed.

Leaves: Simple, alternate or opposite and reduced to firm, sharp spines.

Inflorescence: Flowers solitary in the leaf axils.

> **Calyx:** Sepals 5, slightly united below, falling as a unit, flared above into veiny-translucent wings.
>
> **Corolla:** Petals absent.
>
> **Stamens:** 5 or fewer, minute.
>
> **Pistil:** Ovary superior, style usually with 2 branches.

Fruit: A flattened utricle.

General Comments: Introduced. Russian thistle is common in cultivated and fallow fields; on dry roadsides; and in other dry, disturbed sites. It is conspicuous in the landscape in the arid regions of the Southwest. The mature plants are one of the common tumbleweeds of the western United States. Hart et al. (2003) report that Russian thistles may accumulate toxic levels of nitrates.

CONVOLVULACEAE

Texas Bindweed, Gray Bindweed

Convolvulus equitans
G. Bentham

Growth Habit: Perennial.

Stems: Prostrate, trailing and twining, grayish-green, pubescent; tendrils absent.

Leaves: Simple, alternate; blades variously shaped but usually ovate to lanceolate, densely pubescent, grayish-green, margins lobed; petioles densely pubescent.

Inflorescence: Flowers 1 or 2 at the nodes, pedicels elongated, pubescent.

> **Calyx:** Sepals 5, united near the base, imbricate, pubescent, apex obtuse or rounded.

> **Corolla:** Petals 5, united, actinomorphic, tubular, white or pinkish on the outside and white or red within the throat.

> **Stamens:** 5; epipetalous, anthers light brown.

> **Pistil:** Ovary superior; style 1 with 2 branches.

Fruit: A capsule.

General Comments: Native. Texas bindweed occurs on roadsides and in other disturbed sites. It twines around cultivated plants and interferes with harvests. It is related to the widely distributed weedy plant *C. arvensis* (common bindweed), which has a shorter, nearly glabrous calyx.

Carolina Ponyfoot, Grass Ponyfoot

Dichondra carolinensis
A. Michaux

Growth Habit: Perennial; forming dense mats.

Stems: Prostrate, stoloniferous, rooting at the nodes with a few scattered hairs.

Leaves: Simple, alternate; blades rounded and with palmate venation, pubescent, margins entire; petioles elongated, pubescent.

Inflorescence: Flowers 1 or 2 at the nodes; pedicels elongated, straight and pubescent.

>   **Calyx:** Sepals 5, united, pubescent.

>   **Corolla:** Petals 5, united, light green.

>   **Stamens:** 5; epipetalous.

>   **Pistil:** Ovary superior, deeply lobed, densely pubescent.

Fruit: A 2-lobed, densely pubescent capsule; each lobe with 1 seed.

General Comments: Native. Carolina ponyfoot is a lawn weed that flowers in the spring, but it persists vegetatively for most of the growing season.

Cotton Morningglory, Tievine

Ipomoea cordatotriloba A. Dennstaedt var. *cordatotriloba*

Syn. *I. trichocarpa* S. Elliott

Growth Habit: Perennial; often with milky latex.

Stems: Trailing, twining, climbing; nearly glabrous but with irregularly spaced, pustulate-based hairs.

Leaves: Simple, alternate, slightly pubescent; blades with cordate or auriculate lobes; margins 3 lobed or occasionally entire.

Inflorescence: Flowers 1–2 from the leaf axils; pedicels elongated and with linear bracts.

> **Calyx:** Sepals 5, united below, apex acute.
>
> **Corolla:** Petals 5, united, trumpet or bell shaped, rose to reddish-purple, dark red in the throat.
>
> **Stamens:** 5, epipetalous; filaments pubescent near the base; pollen white.
>
> **Pistil:** Ovary superior; style 1, unbranched.

Fruit: A 4-seeded capsule with a persistent style.

General Comments: Native. Cotton morningglory is a common weed in cotton fields, on fences, and in fallow fields.

Pathogens and Pests: *Ipomoea* species are hosts to an array of plant geminiviruses, including sweet potato leaf curl virus (SPLCV) and *Ipomoea* yellow vein virus (IYVV) (Li, Salih, and Hurtt 2004).

Redthroat Morningglory, Alamo Vine, Cutleaf Morningglory

Merremia dissecta
(N. von Jacquin)
H. Hallier

Syn. *Ipomoea dissecta*
(N. von Jacquin)
F. Pursh; *I. sinuata*
C. Ortega

Growth Habit: Perennial; a vine.

Stems: Twining, pubescent with long hairs arising from reddish pustules.

Leaves: Simple and palmately lobed, alternate; lobes dark green above and light green and veiny on the lower epidermis; petioles elongated with pubescence similar to that of stems.

Inflorescence: Flowers 1 to several from the leaf axils; peduncles elongated and with red pustules.

> **Calyx:** Sepals 5, free and connate and pointed at the apex in the bud stage.
>
> **Corolla:** Petals 5, united below, actinomorphic, white with a reddish-purple throat.
>
> **Stamens:** 5, epipetalous; anthers white.
>
> **Pistil:** Ovary superior; style 1; stigma capitate.

Fruit: A capsule with 4 seeds.

General Comments: Native. This species is a common vine that grows on fences and along canal banks and roadsides.

CRASSULACEAE

Devil's Backbone, Mother of Thousands

Kalanchoë diagremontiana
R. Hamet & H. Perrier

Syn. *Bryophyllum diagremontiana*
(R. Hamet & H. Perrier)
A. Berger

Growth Habit: Perennial; from a shallow, fibrous root system.

Stems: Usually erect, glabrous.

Leaves: Simple, opposite; blades usually broadly lanceolate, glabrous, succulent and mottled, margins serrate and bearing adventitious plantlets.

Inflorescence: A cymose panicle; flowers pendulous.

    Calyx: Sepals 4, succulent.

    Corolla: Petals 4, united and tubular, scarlet or salmon colored.

    Stamens: 8, attached to the midsection of the corolla tube.

    Pistil: Ovary superior.

Fruit: An inconspicuous follicle.

General Comments: Introduced. This invasive species, introduced from Madagascar, is abundant in coastal soils at the Palo Alto Battlefield National Historical Site (Lonard, Richardson, and Richard 2004) and along the Arroyo Colorado in Cameron County (Christina Mild, pers. comm.). Dense carpets of viviparous plantlets form under parent plants.

CUSCUTACEAE

Dodder

Cuscuta spp.

Growth Habit: Annual; parasite of various native species and of cultivated crops; roots absent; chlorophyll apparently absent.

Stems: Forming a tangled, twining mass; orange-yellow, glabrous.

Leaves: Absent.

Inflorescence: A cyme.

    Calyx: Sepals 5, united.

    Corolla: Petals 5, whitish, united and lobed near the apex.

    Stamens: 5, epipetalous.

    Pistil: Ovary superior.

Fruit: A globose capsule.

General Comments: Native. Several closely related species of parasitic dodders occur in South Texas. Flowers and fruits are required for an accurate identification.

EUPHORBIACEAE

Tropical Euphorbia, Graceful Sandmat

Chamaesyce hypericifolia
(C. Linnaeus)
C. Millspaugh

Syn. *Euphorbia hypericifolia* C. Linnaeus

Growth Habit: Annual; from a taproot; plants monoecious, all parts with milky latex.

Stems: Erect, turning red in mature plants, glabrous.

Leaves: Simple, opposite; blades elliptic, glabrous, margins minutely toothed; stipules triangular; petioles present.

Inflorescence: A petaloid cyathium, gland appendages white.

    Staminate Flowers: Numerous staminate flowers per cyathium.

    Pistillate Flowers: Ovary superior; styles 3 each with 2 branches.

Fruit: A glabrous, 3-lobed capsule; seeds less than 1 mm long, brown with lateral ridges.

General Comments: Native. Tropical euphorbia is a year-round weed in flower beds, greenhouses, and agricultural fields.

Prostrate Sandmat

Chamaesyce prostrata
(W. Aiton) J. K. Small

Syn. *Euphorbia prostrata*
W. Aiton

Growth Habit: Annual; from a taproot.

Stems: Prostrate, mat forming but not rooting at the nodes, pubescent.

Leaves: Simple, opposite; blades rounded or oblique at the base, minutely pubescent, margins minutely pubescent, apex rounded; stipules minute; petioles present.

Inflorescence: A cyathium.

> **Staminate Flowers:** 4 glandular staminate groups, flowers with 1 stamen.

> **Pistillate Flowers:** Ovary superior, styles 3 each with 2 branches.

Fruit: A 3-lobed capsule, pubescent on the ridges; seeds brown, about 1 mm long, usually covered with a white substance.

General Comments: Native. This is the most common *Chamaesyce* species in South Texas. It is present in all seasons and is abundant in a variety of disturbed sites. Prostrate sandmat is commonly found in cracks and crevices of sidewalks, where it tolerates extreme temperature fluctuations. It is toxic to livestock if consumed in large amounts (Sperry et al. 1968).

Texas Bullnettle, Mala Mujer

Cnidoscolus texanus
(J. Muller of Aargau)
J. K. Small

Growth Habit: Perennial; from a thick rootstock; milky latex present; plants monoecious.

Stems: Erect, branching; armed with stiff, stinging, glandular-based hairs.

Leaves: Simple, alternate, pubescence similar to that of stems; blades deeply 3–5 lobed, venation palmate; petioles present.

Inflorescence: Terminal, cymose; pistillate flowers few.

> **Calyx:** Sepals 5, petaloid, white, fragrant, densely hispid.

> **Corolla:** Petals absent.

> **Staminate Flowers:** Stamens 10, connate, included within the calyx throat; filaments with soft pubescence at the base.

> **Pistillate Flowers:** Present in the midsection of the inflorescence; ovary superior, hispid; styles 3, each dichotomously branched.

Fruit: A 3-seeded capsule.

General Comments: Native. Texas bullnettle occurs in sandy pastures and on roadsides. It is difficult to control because of its large root system. It causes intense pain when it is touched. Heavy gloves should be worn if the plant is handled.

Lindheimer's Hogwort Croton, Woolly Croton

Croton capitatus
A. Michaux var.
lindheimeri
(G. Engelman &
A. Gray) J. Muller of
Aargau

Growth Habit: Annual; from a taproot; plants monoecious.

Stems: Erect, branching at the nodes; grayish-green; stellate pubescent.

Leaves: Simple, alternate; blades ovate-lanceolate, base cordate to subcordate, silvery pubescent, margins entire; petioles and stipules present.

Inflorescence: A raceme; male flowers located above the females.

Staminate Flowers:

Calyx: Sepals 5, inconspicuous, stellate pubescent.

Corolla: Petals 5, similar to the sepals.

Pistillate Flowers:

Calyx: With 6–9 lobes, stellate pubescent.

Corolla: Petals absent.

Pistil: Ovary superior, densely pubescent; styles 3, lobed.

Fruit: A globose, 3-seeded capsule.

General Comments: Native. This weed dominates sandy, disturbed pastures and range sites during the hot months.

Toothed Spurge, Toothed Poinsettia

Euphorbia dentata
A. Michaux

Syn. *Poinsettia dentata*
(A. Michaux) J. Klotzsch
& C. Garcke

Growth Habit: Annual; from a taproot; all parts with a milky latex; plants monoecious.

Stems: Erect, pubescent.

Leaves: Simple, opposite; blades ovate or ovate-lanceolate, pubescent, light green on the lower epidermis and often with some reddish spots on the uppermost blades, tapered at the base and obtuse at the apex, margins undulate; petioles elongated, pubescent.

Inflorescence: Flowers clustered in terminal cyathia.

> **Staminate Flowers:** Filaments and anthers white.

> **Pistillate Flowers:** Ovary superior, on a short stalk extending from the cyathium; style branches 2, each branch bifid or forked.

Fruit: A glabrous, 3-lobed capsule; seeds slightly mottled.

General Comments: Native. This close relative of poinsettia is an occasional weed in flower beds and greenhouses.

Grassleaf Spurge

Euphorbia graminea
N. J. von Jacquin

Growth Habit: Annual; from a shallow taproot; all parts with milky latex.

Stems: Erect, 4 angled, hollow at the internodes, glabrous, unbranched below but with dichotomous branching above.

Leaves: Simple, opposite; blades ovate, green above and glaucous on the lower epidermis, minutely pubescent; margins entire; petioles elongated, slightly pubescent.

Inflorescence: A cyathium with 2 white, petaloid appendages on an elongated peduncle.

> **Staminate Flowers:** Stamens 5, minute, included within the cyathium; anthers white.

> **Pistillate Flowers:** Ovary superior, on an elongated stalk extending from the cyathium; style branches 3; each bifid.

Fruit: A glabrous, 3-lobed capsule; seeds mottled.

General Comments: Introduced. Mark Mayfield and Hugh Wilson (Texas A&M University) identified this exotic weed. It had previously been reported in Florida, Hawaii, and Mexico. It occurs in greenhouses and is found in flower beds on the campus of the University of Texas–Pan American.

Mascarene Island Leafflower, Tender Leafflower

Phyllanthus tenellus
W. Roxburgh

Growth Habit: Annual; from a taproot; plants monoecious, all parts with a milky latex.

Stems: Erect, glabrous.

Leaves: Simple, alternate; blades ovate-elliptic, glabrous, margins entire; stipules membranous.

Inflorescence: Flowers 1 to several from an elongated peduncle.

 Staminate Flowers:

  Calyx: Sepals 5, glabrous, translucent on the margins.

  Corolla: Petals absent.

  Stamens: 5.

 Pistillate Flowers:

  Calyx: Similar to calyx on staminate flowers.

  Corolla: Absent.

  Pistil: Ovary superior; style branches 5, some bifid.

Fruit: A capsule with each segment bearing 2 brown seeds.

General Comments: Introduced. This exotic species is often present in nurseries and in flower beds.

Castorbean, Higuerilla

Ricinus communis
C. Linnaeus

Growth Habit: Annual; from a taproot; plants monoecious.

Stems: Erect, long-lived, up to 4 m tall or occasionally taller.

Leaves: Simple, alternate; blades palmately veined, glabrous, deeply lobed, margins serrate; petioles elongate, glabrous.

Inflorescence: Female and male flowers in racemose clusters with the males below the females.

Staminate Flowers:

Calyx: Sepals 5, united below, green, glabrous.

Corolla: Petals absent.

Stamens: Numerous, filaments and anthers white.

Pistillate Flowers:

Calyx: Sepals inconspicuous, not persisting.

Corolla: Petals absent.

Pistil: Ovary superior with soft spines, style branches 6, red.

Fruit: A 3-lobed capsule with each lobe bearing 1 seed, seeds brown-and-white mottled (seed resembles a blood-filled tick), caruncle present on the smaller end.

General Comments: Introduced. Castorbeans occur on roadsides, in the Rio Grande riparian zone, and on canal banks. The seeds are a source of castor oil and contain ricin, an agent used in biological warfare. As few as 2 seeds can be lethal to humans. Land developers suggested castorbeans as a cash crop in the LRGV after the Civil War (Dougherty 1869).

FABACEAE

Depressed Wandlike Bundleflower

Desmanthus virgatus
(C. Linnaeus) C. von
Willdenow var. *depressus*
(W. J. Hooker &
R. Barneby *ex*
C. von Willdenow)
B. L. Turner

Growth Habit: Perennial.

Stems: Scandent or sprawling, glabrous, angled.

Leaves: Bipinnately compound, alternate; leaflets glabrous but margins with scattered hairs; stipules present; glands present at base of pinna.

Inflorescence: Few-flowered heads from leaf axils; bracts subtending flowers membranous, brown.

> **Calyx:** Sepals 5, united.

> **Corolla:** Petals 5, actinomorphic, green with white margins.

> **Stamens:** 10; anthers yellow.

> **Pistil:** Ovary superior; style 1, unbranched.

Fruit: A linear legume.

General Comments: Native. This species is present in a wide variety of disturbed sites, including weedy lawns.

Burclover

Medicago polymorpha
C. Linnaeus

Growth Habit: Annual; from a taproot and arising from seeds in previous season's legume; small nodules often present on roots.

Stems: Erect or usually spreading, glabrous.

Leaves: Trifoliolate, alternate; leaflets obovate, glabrous above, minutely appressed pubescent on the lower epidermis, margins toothed; stipules with linear projections; petioles grooved.

Inflorescence: Several flowers arising from the upper leaf axils.

> **Calyx:** Sepals 5, united for about half its length, slightly pubescent, lobes linear.
>
> **Corolla:** Petals 5, yellow, papilionaceous.
>
> **Stamens:** 10, diadelphous.
>
> **Pistil:** Ovary superior; style 1, unbranched.

Fruit: A coiled legume with hooked barbs, burlike.

General Comments: Introduced. Burclover is a common lawn weed during the cool season.

Pathogens and Pests: Although some resistance is present, most accessions of *M. polymorpha* in the *Medicago* core germplasm collection at USDA-ARS, Beltsville, Maryland, are susceptible to *Colleto trichum trifolii* (anthracnose) (O'Neill and Bauchan 2000), *Phoma medicaginis* (spring black stem and leaf spot) (O'Neill and Bauchan 2003), and *Erysiphe pisi* (powdery mildew) (Yaege and Stuteville 2002). However, *M. polymorpha* appears to be highly resistant to *Peronospora trifoliorum* (downy mildew) (Yaege and Stuteville 2000). All of these fungi are important pathogens of alfalfa (*M. sativa*) (Rhodes 2001).

White Sweetclover

Melilotus albus
F. Medicus

Growth Habit: Annual; from a taproot.

Stems: Erect with a few minute hairs, mostly glabrous.

Leaves: Trifoliolate, alternate; leaflets ovate-lanceolate, nearly glabrous, margins serrate; petioles pubescent; stipules present.

Inflorescence: A contracted raceme; pedicels recurved, subtended by a linear bract; flowers with a sweet odor.

    Calyx: Sepals 5, united below with a few scattered hairs.

    Corolla: Petals 5, papilionaceous, white.

    Stamens: 10, diadelphous, anthers yellow.

    Pistil: Ovary superior; style 1.

Fruit: A 1-seeded, podlike legume.

General Comments: Introduced. White sweetclover is an occasional cool-season weed on roadsides and other disturbed sites. It is toxic to cattle and sheep (Sperry et al. 1968; Hart et al. 2003).

Annual Sourclover, Indian Sweetclover

Melilotus indicus
(C. Linnaeus) C. Allioni

Growth Habit: Annual; from a shallow taproot, fragrant.

Stems: Erect, glabrous, green or reddish tinged.

Leaves: Trifoliolate, alternate; leaflets oblanceolate, glabrous above but appressed pubescent on the veins, margins sinuate near the apex; petioles grooved, pubescent; stipules present.

Inflorescence: Flowers in terminal racemes, pedicels recurved.

    Calyx: Sepals 5, united below.

    Corolla: Petals 5, yellow, zygomorphic, papilionaceous.

    Stamens: Diadelphous, in a 9 + 1 arrangement.

    Pistil: Ovary superior; style 1.

Fruit: A 1-seeded, podlike legume.

General Comments: Introduced. This cool-season species is found in sites similar to those where white sweetclover is found but is more common.

Pink Sensitivebrier, Powderpuff

Mimosa strigillosa
J. Torrey & A. Gray

Growth Habit: Perennial; rooting at some nodes; woody near the base; roots with nodules.

Stems: Prostrate, trailing; young stems with coarse, appressed, white hairs and recurved prickles.

Leaves: Sensitive to touch; bipinnately compound, alternate; leaflets pubescent; rachis and petioles pubescent with a swollen pulvinus at the base of leaflet segments and at the base of the petiole; stipules present, reddish.

Inflorescence: Flowers clustered in a globose head; pedicels elongated, pilose.

> **Calyx:** Sepals 5, united below, membranous.
>
> **Corolla:** Petals absent.
>
> **Stamens:** Numerous (more than 10); filaments red or rose colored; anthers white.
>
> **Pistil:** Ovary superior; style 1.

Fruit: An oblong, jointed, pubescent legume.

General Comments: Native. Pink sensitivebrier is a common lawn weed that flowers during the warm season.

White Clover, Dutch Clover

Trifolium repens
C. Linnaeus

Growth Habit: Perennial.

Stems: Prostrate, mat forming, rooting at the nodes; glabrous.

Leaves: Trifoliolate, alternate; leaflets often with a light green patch near the base, glabrous.

Inflorescence: Flowers in a head on an elongated, glabrous peduncle; pedicels subtended by a small, white bract.

> **Calyx:** Sepals 5, united, white with greenish, linear lobes, glabrous.
>
> **Corolla:** Petals 5, white, turning pink with age, papilionaceous.
>
> **Stamens:** 10, diadelphous; anthers yellow.
>
> **Pistil:** Ovary superior; style 1, unbranched.

Fruit: A few-seeded loment.

General Comments: Introduced. Several states have listed white clover as an invasive species. It is a lawn weed that flowers during the cool season.

Pathogens and Pests: Species of *Trifolium* are susceptible to clover yellow mosaic virus (ClYMV), clover yellow vein virus (ClYVV), and bean yellow mosaic virus (BYMV).

Louisiana Vetch, Slim Vetch, Deer Pea Vetch

Vicia ludoviciana
T. Nuttall subsp.
ludoviciana

Growth Habit: Annual.

Stems: Sprawling, ribbed, minutely pubescent.

Leaves: Pinnately compound, terminal leaflet reduced to a tendril, alternate; leaflets glabrous above and appressed pubescent on the lower epidermis.

Inflorescence: Flowers in axillary racemes; peduncles elongated, slightly pubescent; stipules present.

> **Calyx:** Sepals 5, united, pubescent, lobes linear.
>
> **Corolla:** Petals 5, bluish-purple, papilionaceous.
>
> **Stamens:** 10, diadelphous.
>
> **Pistil:** Ovary superior, style 1, unbranched.

Fruit: A laterally compressed legume (resembles a small snow pea).

General Comments: Native. Louisiana vetch is of minor importance as a cool-season lawn weed. It disappears during the warm season.

HALORAGACEAE

Eurasian Watermilfoil

Myriophyllum sibiricum V. L. Komarov

Syn. *M. spicatum* C. Linnaeus var.
exalbescens (M. Fernald) W. Jepson

Growth Habit: Perennial; submersed
aquatic; rooted in bottom mud; plants
monoecious.

Stems: Brown, reddish-brown or
purplish; up to 2–3 m long.

Leaves: Simple and pinnately dissected,
whorled, usually about 1 cm apart on
the stem; leaf segments linear (leaf
resembles a damaged feather).

Inflorescence: A spike; lower flowers
pistillate; upper flowers staminate.

> **Staminate Flowers:** Calyx with 4
> segments; stamens 8.

> **Pistillate Flowers:** Corolla of 4
> small petals; stigmas 4.

Fruit: A 4-lobed nutlet.

General Comments: Introduced. Eurasian watermilfoil is native to Europe and
Asia but is now found in many areas of the world, including the United States and
Canada. This invasive aquatic weed occurs in Lake Amistad and the Rio Grande
near Del Rio, Texas, and in Coleto Creek Reservoir near Goliad, Texas. Eurasian
watermilfoil will displace other submerged plant species, reducing both habitat
diversity and plant species diversity (Nichols and Shaw 1986). If not controlled, it
can also prohibit aquatic recreation. Triploid grass carp have been used successfully to control this aquatic weed. The most efficient herbicide used on Eurasian
watermilfoil is 2,4-D (Anonymous 2001).

HYDROPHYLLACEAE

Jamaican Nama

Nama jamaicense
C. Linnaeus

Growth Habit: Annual; from a taproot.

Stems: Prostrate, pubescent.

Leaves: Simple, alternate; blades spatulate, pubescent, margins entire; petioles obscure but forming a wing on the stem to the node below.

Inflorescence: Flowers 1 to several from the leaf axils; pedicles pubescent.

    Calyx: Sepals 5, united below, pubescent.

    Corolla: Petals 5, united for most of the length, white.

    Stamens: 5; epipetalous, anthers cream colored.

    Pistil: Ovary superior, style bifurcate at the apex.

Fruit: A capsule with numerous seeds.

General Comments: Native. Jamaican nama occurs occasionally in lawns and flower beds.

Smallleaf Nama

Nama parvifolium
(J. Torrey) J. Greenman

Growth Habit: Annual; from a taproot.

Stems: Prostrate, mat forming, rooting at the nodes, pubescent with gland-tipped hairs.

Leaves: Simple and clustered below and opposite above; blades ovate to lanceolate, pubescent with gland-tipped hairs, margins entire; petioles pubescent.

Inflorescence: Flowers solitary in the leaf axils.

> **Calyx:** Sepals 5, united at the base, exceeding the length of the corolla, pubescent.
>
> **Corolla:** Petals 5, united into a tube below, purple above and light green in the tube.
>
> **Stamens:** 5, epipetalous.
>
> **Pistil:** Ovary superior; style 1.

Fruit: A capsule with numerous seeds.

General Comments: Native. Smallleaf nama is a cool-season weed in agricultural fields. It disappears during the warm season.

LAMIACEAE (LABIATAE)

Henbit

Lamium amplexicaule
C. Linnaeus

Growth Habit: Annual; from a taproot.

Stems: Creeping, decumbent, or erect, 4 angled, minutely pubescent.

Leaves: Simple, opposite, lower leaves petiolate, upper leaves (bracts) clasping; blades rounded, pubescent, margins toothed.

Inflorescence: Flowers subtended by rounded, leafy bracts that resemble leaves; flowers in dense, axillary spikes.

> **Calyx:** Sepals 5, united below, pubescent.
>
> **Corolla:** Petals united and forming a bilabiate structure, reddish-purple or pink with nectar guides on the lateral appendages, pubescent.
>
> **Stamens:** 4, epipetalous; anthers pubescent; pollen orange.
>
> **Pistil:** Ovary superior, 4 lobed.

Fruit: 4 bony, mottled nutlets.

General Comments: Introduced. Henbit is a cool-season weed in lawns, citrus groves, and fields. It is much more common in the temperate zone.

Pathogens and Pests: *Lamium amplexicaule* may serve as a host for *Sclerotinia minor,* a fungus causing sclerotinia blight of peanuts and other dicots (Hollowell et al. 2003).

MALVACEAE

Little Mallow

Malva parviflora
C. Linnaeus

Growth Habit: Annual; from a taproot.

Stems: Absent or a small basal crown.

Leaves: In a cluster arising from the crown, simple; blades palmately veined, stellate pubescent, margins shallowly lobed and with blunt teeth; petioles elongated, stellate pubescent.

Inflorescence: Flowers in axillary and terminal clusters.

> **Calyx:** Subtended by linear, pubescent involucral bracts (involucel); sepals 5, united below, pubescent primarily on the margins.

> **Corolla:** Petals 5, free above but united on a disk below, cleft at the apex, white and pinkish-purple at the apex.

> **Stamens:** Monadelphous, numerous; pollen white.

> **Pistil:** Ovary superior.

Fruit: A disklike capsule, pubescent, segments 10, bearing 1 seed in each segment.

General Comments: Introduced. Little mallow is common on dry roadsides during the cool season.

Pathogens and Pests: Idris et al. (2003) have shown that *M. parviflora* is a host for *Macroptillon* yellow mosaic Florida virus (MaYMFV), a begomovirus vectored by whitefly (*Bemesia tabaci*) that may cause disease in the common bean (*Phaseolus vulgaris*).

American Falsemallow, Indian Valley Falsemallow, Rio Grande Falsemallow

Malvastrum americanum
(C. Linnaeus) J. Torrey

Growth Habit: Annual; from a taproot, robust.

Stems: Erect, stellate pubescent, red to reddish-green.

Leaves: Simple, alternate; blades ovate to rounded, stellate pubescent, margins broadly toothed (crenate); petioles stellate pubescent; stipules linear, pubescent.

Inflorescence: In few-flowered axillary clusters and terminal spicate racemes.

> **Calyx:** Subtended by linear, appressed pubescent involucral bracts (involucel); sepals 5, united, pubescent and with reddish papillose-based marginal hairs.
>
> **Corolla:** Petals 5, united near the base, yellow.
>
> **Stamens:** Monadelphous, numerous, pollen yellow.
>
> **Pistil:** Ovary superior.

Fruit: A disklike, segmented, pubescent capsule.

General Comments: Native. *Malvastrum americanum* is a pantropical weed common in old fields and in other disturbed sites. Heavy grazing followed by fire appears to stimulate population growth.

Threelobe Falsemallow

Malvastrum coromandelianum
(C. Linnaeus) C. Garcke

Growth Habit: Annual or perennial.

Stems: Erect or usually sprawling, appressed stellate pubescent.

Leaves: Simple, alternate; blades ovate or broadly elliptic, stellate pubescent, margins crenate, apex obtuse or occasionally acute; stipules linear; petioles stellate pubescent.

Inflorescence: Flowers solitary in the leaf axils.

> **Calyx:** Subtended by an involucel of 3 linear, pubescent bracts; sepals 5, united for 1/3 to 1/2 their length, pubescent, acute at the apex.

> **Corolla:** Petals 5, united near the base, yellow, opening in late afternoon.

> **Stamens:** Numerous, monadelphous; anthers yellow.

> **Pistil:** Ovary superior.

Fruit: A pubescent capsule separating into numerous 1-seeded segments; each segment with lateral bristles and 2 hornlike spines.

General Comments: Native. Threelobe falsemallow is one of the most difficult dicot weeds to eradicate in lawns. It competes successfully with bermudagrass and St. Augustine grass (*Stenotaphrum secundatum*).

Bristly Mallow

Modiola caroliniana
(C. Linnaeus) G. Don

Growth Habit: Perennial; rooting at the nodes.

Stems: Prostrate with scattered bristly, pustulate-based hairs.

Leaves: Simple, alternate; blades round, cordate at the base, nearly glabrous above but pubescent on the lower epidermis, venation palmate, margins lobed and bluntly toothed; petioles grooved and densely pubescent; stipules present.

Inflorescence: Flowers 1 to several from the upper leaf axils.

> **Calyx:** Subtended by an involucel of 3 spatulate, pubescent bracts; sepals 5, united at the base.
>
> **Corolla:** Petals 5–7, united near the base, salmon colored.
>
> **Stamens:** Numerous, monadelphous; anthers yellow.
>
> **Pistil:** Ovary superior, pubescent, style branches red.

Fruit: A capsule.

General Comments: Native. *Modiola caroliniana* is uncommon in South Texas, but it is present in bermudagrass turf on the campus of the University of Texas–Pan American. It is more common as a lawn weed in central and southeastern Texas.

Tuberous Sida

Rhynchosida physocalyx
(A. Gray) P. Fryxell

Syn. *Sida physocalyx*
A. Gray,
non F. von Mueller

Growth Habit: Perennial.

Stems: Trailing or sprawling, stellate pubescent.

Leaves: Simple, alternate; blades broadly ovate, venation palmate, stellate pubescent, margins toothed; petioles elongated, stellate pubescent; stipules linear, pubescent.

Inflorescence: Flowers solitary from the leaf axils, pedicels stellate pubescent.

>   **Calyx:** Sepals 5, united below, pubescent, enlarged and persistent in fruit.

>   **Corolla:** Petals 5, united below, light yellow, actinomorphic.

>   **Stamens:** Numerous, monadelphous, anthers yellow.

>   **Pistil:** Ovary superior with a mealy surface and 10 lobes.

Fruit: A lobed capsule.

General Comments: Native. This weedy species occurs on roadsides and in a variety of disturbed sites.

Prickly Sida, Prickly Fanpetals, Teaweed

Sida spinosa C. Linnaeus

Growth Habit: Annual; from a taproot.

Stems: Erect, branching, stellate pubescent.

Leaves: Simple, alternate; blades linear-lanceolate, stellate pubescent, margins crenate; stipules linear, pubescent; petioles with a blunt spine near the base.

Inflorescence: Flowers usually 1 or occasionally several arising from the leaf axils.

    Calyx: Sepals 5, united, margins often purple tinged, stellate pubescent.

    Corolla: Petals 5, united near the base, actinomorphic, yellow.

    Stamens: Numerous, monadelphous; anthers yellow.

    Pistil: Ovary superior, styles numerous.

Fruit: The calyx enlarging and enclosing a capsule; capsule with spine-tipped carpel segments.

General Comments: Native. Prickly sida occurs on roadsides, in cultivated and fallow fields, and in other disturbed sites.

MOLLUGINACEAE

Carpetweed, Indian Chickweed

Mollugo verticillata
C. Linnaeus

Growth Habit: Annual; from a small taproot.

Stems: Prostrate, branching freely, glabrous.

Leaves: Simple, whorled; blades linear, glabrous, margins entire.

Inflorescence: Few flowered axillary clusters; pedicels elongated.

> **Calyx:** Sepals 5, free, green on the back and white on the inner surface, petaloid, glabrous.
>
> **Corolla:** Petals absent.
>
> **Stamens:** 3 or 4; filaments white.
>
> **Pistil:** Ovary superior, glabrous; styles 3.

Fruit: A lobed, many-seeded capsule; seeds brown, resembling a snail.

General Comments: Introduced. Carpetweed is usually found in moist, bare, sandy or sandy loam soils. It is a common garden weed in southeastern Texas.

MORACEAE

Mulberry Weed, Shaggy Crabweed

Fatoua villosa
(C. Thunberg)
T. Nakai

Growth Habit: Annual; from a taproot; plants monoecious.

Stems: Erect, pubescent with some hairs straight and gland tipped and others curved or bent.

Leaves: Simple, alternate; blades broadly ovate, nearly glabrous above and pubescent on the lower epidermis along the veins, margins shallowly lobed, apex obtuse or acute; petioles elongated and pubescent.

Inflorescence: Flowers unisexual, in axillary clusters; inflorescences distributed along the entire stem axis.

>   **Calyx:** Sepals 4, partially united below, pubescent with glandular-tipped hairs, reddish-purple above (similar in males and females).

>   **Corolla:** Absent.

>   **Stamens:** 4; anthers white, filaments short.

>   **Pistil:** Ovary superior; style 1, unbranched, purple tinged.

Fruit: A utricle or an achene with an angular, mottled seed.

General Comments: Introduced. Mulberry weed, introduced from Asia, is listed as a noxious species in Alabama and Tennessee. In South Texas, it is occasionally found in flower beds or in flowerpots in nurseries.

NYCTAGINACEAE

Red Spiderling

Boerhavia coccinea
P. Miller

Growth Habit: Perennial.

Stems: Prostrate and spreading, mat forming, minutely pubescent.

Leaves: Simple, opposite; blades ovate, unequal, green above and silver-gray on the lower epidermis, margins pubescent, undulate; petioles pubescent.

Inflorescence: Flowering from dense, axillary clusters; peduncle pubescent.

    Calyx: Sepals 5, united, petaloid, crimson.

    Corolla: Absent.

    Stamens: 1 to several.

    Pistil: Ovary superior, covered with viscid glands.

Fruit: A ribbed, viscid achene.

General Comments: Native. Red spiderling is a common weed in dry lawns. It is difficult to eradicate by hand.

Erect Spiderling

Boerhavia erecta
C. Linnaeus

Growth Habit: Annual; from a taproot.

Stems: Erect, glabrous, often red below.

Leaves: Simple, opposite; blades ovate, dark green and minutely papillose above, silver-gray below and minutely papillose on the lower epidermis, margins undulate; petioles pubescent and often reddish tinged.

Inflorescence: Terminal and axillary paniculate cymes.

    Calyx: Sepals 5, united and each with 2 lobes, pink, petaloid.

    Corolla: Absent.

    Stamens: 1–4; anthers white.

    Pistil: Ovary superior; style 1, unbranched.

Fruit: A glabrous, club-shaped, 5-ribbed achene.

General Comments: Native. Erect spiderling is a common weed in flower beds and on the margins of highways.

ONAGRACEAE

Plains Beeblossom

Gaura brachycarpa
J. K. Small

Growth Habit: Annual.

Stems: Sprawling and arching upward near the apex, pubescent.

Leaves: Simple, alternate; blades lanceolate, pubescent, margins sinuate.

Inflorescence: A remotely flowered spike; flowers subtended by a small bract.

>   **Calyx:** Sepals 4, attached to the apex of a hypanthium, free, green to reddish-pink and recurved, pubescent.
>
>   **Corolla:** Petals 4, free, stalked, slightly zygomorphic, reddish-pink.
>
>   **Stamens:** 8; filaments white; anthers red; pollen white.
>
>   **Pistil:** Ovary superior, 4 lobed; style 1.

Fruit: A sessile, pyramid-shaped, 4-lobed, pubescent capsule.

General Comments: Native. Plains beeblossom is common on sandy roadsides. It is similar to *G. drummondii* (E. Spach) J. Torrey & A. Gray (Drummond's beeblossom), which has slightly longer petals and larger capsules.

Smallflower Gaura

Gaura parviflora
D. Douglas *ex*
J. G. Lehmann

Growth Habit: Annual; from an enlarged taproot.

Stems: Erect, pubescent with long, soft hairs and shorter soft hairs.

Leaves: Simple, alternate; blades ovate-lanceolate with a soft, velvety texture, margins remotely serrate, pubescent; petioles pubescent with young leaves at the nodes.

Inflorescence: An elongated spike; flowers arising from knoblike extensions from the rachis and subtended by ephemeral, linear, pubescent bracts.

> **Calyx:** Sepals 4, attached to the apex of a hypanthium, green to reddish-pink, glabrous.
>
> **Corolla:** Petals 4, free, white or light pink.
>
> **Stamens:** 8; filaments white; anthers red; pollen white.
>
> **Pistil:** Ovary inferior, 4 lobed; style 1.

Fruit: A glabrous capsule tapered slightly at the apex and near the base.

General Comments: Native. *Gaura parviflora* is a tall, weedy species that often forms large stands in a wide variety of disturbed sites, including roadsides and fallow fields.

Cutleaf Eveningprimrose

Oenothera laciniata
J. Hill

Growth Habit: Perennial.

Stems: Prostrate or decumbent, pubescent.

Leaves: Simple, alternate; blades pinnately lobed, pubescent; petioles absent.

Inflorescence: Flowers solitary from the leaf axils.

> **Calyx:** Sepals 4, pubescent, attached at the apex of a floral tube (hypanthium).

> **Corolla:** Petals 4, yellow, attached at the apex of a floral tube.

> **Stamens:** 8.

> **Pistil:** Ovary inferior; style elongated, 4 lobed at the apex.

Fruit: A cylindrical, pubescent capsule, scarcely tapered at the base; seeds numerous.

General Comments: Native. Cutleaf eveningprimrose is common in flower beds and in sandy and cultivated fields. The flowers open at dusk.

Pathogens and Pests: *Oenothera laciniata* may serve as a host for *Sclerotinia minor,* a fungus causing sclerotinia blight of peanuts and other dicots (Hollowell et al. 2003).

Showy Eveningprimrose, Pink Eveningprimrose, Amapola Del Campo

Oenothera speciosa
T. Nuttall

Growth Habit: Perennial.

Stems: Erect with long, soft hairs.

Leaves: Simple, alternate; blades ovate, mostly glabrous but pubescent on the remotely sinuate margins; petioles pubescent.

Inflorescence: Flowers 1 to several from the upper leaf axils; pedicels absent (seeds are present in a structure that resembles a pedicel).

> **Calyx:** Sepals 4, united longitudinally with reddish margins, glabrous on the upper surface and pubescent on the outer epidermis; calyx and corolla attached to the apex of an elongated hypanthium.

> **Corolla:** Petals 4, free, pink with prominent nectar guides, greenish-white near the base.

> **Stamens:** 8; filaments attached at the midsection of the anthers; pollen cream colored.

> **Pistil:** Ovary inferior, pubescent; style branched into 4 segments, elevated slightly above the stamens.

Fruit: A 4-angled, club-shaped, pubescent capsule.

General Comments: Native. This species is one of the most attractive cool-season wildflowers, but it may be a weed in lawns, flower beds, and fallow fields.

OXALIDACEAE

Drummond's Woodsorrel

Oxalis drummondii A. Gray

Growth Habit: Perennial; from a bulb.

Stems: Aerial stems absent.

Leaves: Arising from a bulb, trifoliolate, leaflets resembling a "kite" in outline, glabrous, margins entire; petioles elongated.

Inflorescence: A few-flowered cluster, axis glabrous.

> **Calyx:** Sepals 5, united with an orange or brown patch at the apex.
>
> **Corolla:** Petals 5, purple, united near the base, filaments with glandular hairs.
>
> **Stamens:** 10, monadelphous, in 2 series of 5 stamens each.
>
> **Pistil:** Ovary superior, lobed; styles 5.

Fruit: A lobed capsule with numerous round seeds.

General Comments: Native. Drummond's woodsorrel often forms dense patches in lawns and flower beds during the cool season. It is difficult to eradicate.

Pathogens and Pests: Several species of *Oxalis* (including *O. corniculata* and *O. drummondii*) serve as aecial hosts for the rust fungus *Puccinia sorghi,* which causes common rust of sorghum and corn. The aecia pustules represent the "sexual reproduction" stage of the rust fungus. These pustules are bright yellow in color and are located on the lower epidermis of the leaflets. The aeciospores rub off onto the fingers readily if the pustules are touched. The spermatogonial stage occurs on the upper epidermis of the leaflets but is generally not remarkable.

Common Yellow Woodsorrel, Yellow Sheepsorrel

Oxalis stricta C. Linnaeus

Growth Habit: Annual or perennial.

Stems: Creeping or sprawling, pubescent.

Leaves: Palmately trifoliolate, alternate; leaflets obcordate, pubescent, margins entire, apex rounded and cleft; stipules linear, pubescent; petioles pubescent.

Inflorescence: Racemose or paniculate; pedicels pubescent.

    Calyx: Sepals 5, loosely united at the base, yellow.

    Corolla: Petals 5, loosely united at the base, yellow.

    Stamens: 10, monadelphous in 2 series of 5 stamens each.

    Pistil: Ovary superior; styles 5.

Fruit: An elongated, pubescent capsule; seeds amber.

General Comments: Native. This species is a weed in flower beds and in other disturbed sites. Microscopic crystals of oxalic acid are present in the leaflets.

PAPAVERACEAE

Golden Pricklypoppy, Bronze Pricklypoppy

Argemone aenea
G. Ownbey

Growth Habit: Annual; from a taproot; all parts with a yellow latex.

Stems: Erect with stout, perpendicular prickles that are painful to the touch.

Leaves: Simple, alternate; blades lanceolate, clasping, primary veins outlined with broad white lines, lobed, with numerous perpendicular prickles.

Inflorescence: Flowers solitary from the apex and from subterminal leaf axils.

    Calyx: Sepals 3, spiny (buds are spiny and have 3 hornlike spines at the apex).

    Corolla: Petals 5 or 6, large, fragile, free, yellow or bronze and showy.

    Stamens: 100–150; filaments red above and yellow at the base.

    Pistil: Ovary superior, spinose; stigma capitate and bearing several pits.

Fruit: A spiny capsule with numerous seeds.

General Comments: Native. This is an attractive cool-season weed that is painful to the touch.

Mexican Pricklypoppy, Yellow Pricklypoppy, Devil's Fig, Chardo Santo, Chicalote

Argemone mexicana
C. Linnaeus

Growth Habit: Annual; from a taproot; all parts with a yellow latex.

Stems: Erect with stout, perpendicular prickles.

Leaves: Simple, alternate; blades lanceolate with numerous prickles, clasping, margins lobed.

Inflorescence: Flowers solitary from the apex and from subterminal leaf axils.

> **Calyx:** Sepals 3, spiny in the bud stage.
>
> **Corolla:** Petals 5 or 6, free, bright yellow, fragile.
>
> **Stamens:** 30–50.
>
> **Pistil:** Ovary superior, spinose; stigma capitate.

Fruit: A spiny capsule with numerous seeds.

General Comments: Native. Mexican pricklypoppy is a cool-season weed that is usually present on sandy roadsides and in sandy pastures.

Spiny Pricklypoppy, Red Pricklypoppy

Argemone sanguinea
E. Greene

Growth Habit: Annual; from a taproot, all parts with a viscous, yellow latex.

Stems: Erect, bluish-green with firm spines that are painful to the touch.

Leaves: Simple, alternate; blades pinnatisect, bluish-green, spiny, white on the upper epidermis over the primary veins, margins spine tipped; petioles present on lower leaves, subsessile on the upper leaves.

Inflorescence: Flowers in a terminal panicle with spiny bracts subtending the calyx; flowers large.

> **Calyx:** Sepals 2–3 with spiny, hornlike crests.
>
> **Corolla:** Petals 5, free, often wrinkled, white or red, actinomorphic, not persisting.
>
> **Stamens:** Numerous, many more than 10; filaments and anthers yellow.
>
> **Pistil:** Ovary superior, spiny; style 1 with a reddish stigmatic surface that resembles a starfish.

Fruit: A spiny capsule; seeds brown with a warty surface.

General Comments: Native. Spiny pricklypoppy is abundant during the cool season and occasionally in the summer on railroad rights-of-way and roadsides.

PLANTAGINACEAE

Hooker's Plantain, California Plantain, Tallowweed

Plantago hookeriana
F. von Fischer &
C. von Meyer

Growth Habit: Annual; from a shallow taproot.

Stems: Absent, but the leafless, pubescent scape resembles a stem.

Leaves: Simple, in a basal rosette; blades linear, gradually tapering to a petiole, appressed pubescent, grayish-green, margins entire.

Inflorescence: A terminal spike with pubescent bracts subtending each flower; bracts shorter than the calyx with green centers and translucent margins.

> **Calyx:** Sepals 4, free, pubescent with translucent margins.
>
> **Corolla:** Petals 4, connate, lobes spreading, translucent with lavender spots in the throat.
>
> **Stamens:** 4, epipetalous; filaments exserted.
>
> **Pistil:** Ovary superior, pubescent; style 1.

Fruit: A capsule.

General Comments: Native. Hooker's plantain is present in deep, sandy soils during the cool season.

Red Seed Plantain

Plantago rhodosperma
J. Decaisne

Growth Habit: Annual; from a taproot.

Stems: Absent but with a scape.

Leaves: Simple in a basal rosette; blades oblanceolate, obtuse at the apex, pubescent, margins irregularly toothed; petioles pubescent on the margins.

Inflorescence: Flowers in a linear, elongated spike; scape densely pubescent, leafless.

>**Calyx:** Subtended by keeled, pubescent bract(s); sepals 4, pubescent, translucent on the margins.

>**Corolla:** Petals 4, united, conelike, whitish or light brown.

>**Stamens:** 4, epipetalous.

>**Pistil:** Ovary superior; style 1.

Fruit: A circumscissile capsule; seeds reddish.

General Comments: Native. *Plantago rhodosperma* is found in a wide variety of disturbed sites during the cool season.

POLYGONACEAE

Heartsepal Buckwheat, Manyflowered Buckwheat

Eriogonum multiflorum
G. Bentham

Growth Habit: Annual; from a taproot.

Stems: Erect with tomentose or densely packed, appressed white hairs.

Leaves: Simple, alternate; blades ovate, sessile, light green and tomentose above and the lower epidermis densely pubescent with white hairs, margins undulate.

Inflorescence: Paniculate-corymbose; short bracts present at base of peduncle.

> **Calyx:** Sepals 6, winged, stipitate, white with a dark green midvein.
>
> **Corolla:** Absent.
>
> **Stamens:** 9, free, densely pubescent on the lower portion of the filaments.
>
> **Pistil:** Ovary superior.

Fruit: An achene.

General Comments: Native. This species forms large stands in sandy pastures and on roadsides during the fall.

Amnastla Dock, Goldenfruited Dock

Rumex chrysocarpus
G. Moris

Growth Habit: Perennial.

Stems: Usually erect, often reddish-purple below.

Leaves: Simple, crowded in a basal rosette, alternate above; blades lanceolate or oblong, glabrous, margins sinuate; petioles elongated and surrounding the stem as a scaly ocrea.

Inflorescence: Flowers in dense clusters at the nodes.

> **Calyx:** Sepals 3, free with an inflated tubercle in the midsection and lateral spines; tightly enclosing the pistil.
>
> **Corolla:** Petals absent.
>
> **Stamens:** 6 or 9.
>
> **Pistil:** Ovary superior; style 1.

Fruit: A triangular achene.

General Comments: Native. *Rumex chrysocarpus* is often abundant in heavy, wet, clay soils in ditches, canal overflows, and freshwater wetlands during the cool season. Hart et al. (2003) indicate that all species of dock may contain toxic levels of oxalates.

PORTULACACEAE

Common Purslane, Verdolaga

Portulaca oleracea
C. Linnaeus

Growth Habit: Annual; from a taproot.

Stems: Erect or prostrate, succulent, glabrous.

Leaves: Simple, opposite or whorled near the stem apex; blades ovate, succulent, margins entire, apex rounded, petioles glabrous.

Inflorescence: Flowers in terminal clusters.

> **Calyx:** Sepals 2, cone shaped, attached to a hypanthium.
>
> **Corolla:** Petals 5, slightly united near the base, attached to a hypanthium, yellow.
>
> **Stamens:** 10, free; anthers yellow.
>
> **Pistil:** Ovary partially inferior, perigynous.

Fruit: A many-seeded capsule; seeds mottled, black.

General Comments: Native. Common purslane occurs in flower beds, gardens, lawns, agricultural fields, and cracks and crevices in pavement. Some authors interpret the floral arrangement as consisting of an involucre of 2 bracts and petaloid tepals.

RANUNCULACEAE

Drummond's Virginsbower, Texas Virginsbower, Old Man's Beard, Barbas De Chivato

Clematis drummondii
J. Torrey & A. Gray

Growth Habit: Perennial.

Stems: A ribbed, pubescent vine.

Leaves: Compound with 3–7 variously lobed leaflets; blades opposite, pubescent; uppermost leaves simple; petiolules twining.

Inflorescence: Flowers solitary from the leaf axils on elongated, ribbed, pubescent pedicels; plants with bisexual, monoecious, or occasionally dioecious flowers.

> **Calyx:** Sepals 4, free, whitish at maturity, pubescent.

> **Corolla:** Petals absent.

> **Stamens:** Numerous, more than 10 per flower; filaments white, flattened; anthers brown.

> **Pistil:** Flowers with numerous separate pistils; ovary superior, styles unbranched, plumose.

Fruit: Achenes numerous; styles plumose (silky), white, persistent.

General Comments: Native. *Clematis drummondii* is a vinelike species that forms tangles in citrus groves, on fences, and in fallow fields. It is difficult to eradicate. It causes skin irritations in susceptible people.

Plains Larkspur

Delphinium carolinianum
T. Walter subsp.
virescens (T. Nuttall).
C. J. Brooks

Syn. *D. virescens*
T. Nuttall var.
macroseratilis
(P. Rydberg) V. Cory

Growth Habit: Perennial; from deep, somewhat woody roots.

Stems: Erect, unbranched, pubescent.

Leaves: Simple, alternate; blades palmately cleft, dissected to the midrib, pubescent.

Inflorescence: A spicate raceme.

> **Calyx:** Sepals 5, the upper sepal zygomorphic and spurlike, resembling a petal.
>
> **Corolla:** Petals 4, zygomorphic, the upper pair tapered at the base into a spur, white or light blue.
>
> **Stamens:** Numerous, more than 10.
>
> **Pistils:** 3; ovary superior.

Fruit: A follicle with numerous seeds.

General Comments: Native. Plains larkspur is poisonous to livestock (Sperry et al. 1968). Cattle are more sensitive than sheep and goats (Hart et al. 2003). This attractive species is usually found in deep, sandy soils during the cool season.

RUBIACEAE

Poorjoe, Rough Buttonweed

Diodia teres T. Walter

Growth Habit: Annual; from a taproot.

Stems: Branching above the base, pubescent with hairs of different lengths, often purple tinged.

Leaves: Simple, opposite; blades linear to narrowly lanceolate, scabrous, margins with minute teeth; stipular bristles numerous, the larger pink to purple.

Inflorescence: Flowers 1–3 in the leaf axils.

> **Calyx:** Sepals 4, free, unequal, persistent, the smaller rounded or ovate, the longer lanceolate and bearing a short awn, margins minutely ciliate.

> **Corolla:** Petals 3 or 4 lobed, tubular at the base, white or pinkish, ciliate on the lobes.

> **Stamens:** 4, epipetalous; anthers white.

> **Pistil:** Ovary inferior, ciliate; style 1, unbranched; stigma capitate.

Fruit: A few-seeded capsule subtended by a persistent calyx.

General Comments: Native. Poorjoe is a common species in low-fertility sandy soils.

Prairie Mexican Clover, Threeseeded Mexican Clover

Richardia tricocca
(J. Torrey & A. Gray)
P. Standley

Syn. *Diodia tricocca*
J. Torrey & A. Gray

Growth Habit: Perennial; with scaly rhizomes.

Stems: Prostrate, mat forming, hirsute.

Leaves: Simple, opposite; blades lanceolate, sessile, glabrous, margins entire and with regularly spaced hairs; stipular bristles present.

Inflorescence: Flowers in small clusters in the upper leaf axils.

> **Calyx:** Sepals 4 with marginal bristles, united near the base.

> **Corolla:** Petals 3 or 4, united, white, actinomorphic.

> **Stamens:** 4, epipetalous, attached to the base of the corolla lobes; pollen white.

> **Pistil:** Ovary inferior, bristly.

Fruit: A bristly capsule with a persistent calyx; seeds 3 or 4.

General Comments: Native. This species often co-occurs with rough buttonweed in sandy, disturbed sites.

SOLANACEAE

Small Groundcherry

Chamaesaracha coronopus (M. Dunal) A. Gray

Growth Habit: Perennial; forming colonies.

Stems: Low growing or prostrate, pubescent with branched hairs; hairs of 2 lengths.

Leaves: Simple, opposite or alternate; blades lanceolate with a broad midvein, pubescence similar to that of stems, margins undulate; petioles pubescent.

Inflorescence: Flowers usually solitary from the leaf axils; pedicels elongated, pubescent.

> **Calyx:** Sepals 5, united, with dense, stiff hairs.
>
> **Corolla:** Petals 5, united below, light yellow and purple tinged with soft, velvety hairs at the base within the corolla.
>
> **Stamens:** 5, epipetalous; anthers yellow.
>
> **Pistil:** Ovary superior; style 1.

Fruit: A round berry.

General Comments: Native. This species occurs on dry, compacted soils. It is closely related to *C. sordida* (M. Dunal) A. Gray (green false nightshade, crowned false nightshade), which has unbranched hairs.

Netted Globeberry

Margaranthus solanaceus
D. von Schlechtendal

Growth Habit: Annual; from a taproot.

Stems: Erect with elongated ribs, minutely pubescent.

Leaves: Simple, alternate; blades ovate or obovate, glabrous, venation pinnate on smaller leaves and palmate on larger leaves, margins undulate; petioles grooved.

Inflorescence: Flowers solitary in leaf axils; pedicels pubescent.

> **Calyx:** Sepals 5, united, pubescent.
>
> **Corolla:** Petals 5, united (urceolate), purple below, glabrous externally, pubescent internally.
>
> **Stamens:** 5, epipetalous.
>
> **Pistil:** Ovary superior; style 1, unbranched.

Fruit: A berry enclosed by an enlarged, inflated calyx.

General Comments: Native. Netted globeberry occurs on the margins of recently cultivated fields. The flowers resemble miniature Chinese paper lanterns.

Fiddleleaf Tobacco, Star Tobacco

Nicotiana repanda
C. von Willdenow *ex*
J. G. Lehmann

Growth Habit: Annual; from a taproot.

Stems: Erect, pubescent.

Leaves: Simple, alternate; blades repand or ovate, clasping the stem, pubescent; petioles absent on the upper leaves (basal leaves often forming a rosette of much larger leaves).

Inflorescence: A few-flowered raceme.

> **Calyx:** Sepals 5, united, ribbed, lobes lanceolate.

> **Corolla:** Petals 5, funnel-form, white, the tube typically 4–5 cm long, opening in the late afternoon or under cloudy conditions early in the day.

> **Stamens:** 5, epipetalous, attached to the corolla tube; pollen abundant, white.

> **Pistil:** Ovary superior; style 1, clavate.

Fruit: A capsule with numerous seeds.

General Comments: Native. This close relative of tobacco has leaves that resemble small tobacco leaves. The flowers open at dusk.

Pathogens and Pests: Fiddleleaf tobacco is a source of inoculum for *Peronospora tabacina* (blue mold of tobacco). This oomycete overwinters on wild tobacco species below the 30th parallel (David Lemke, pers. comm.).

Smallflower Groundcherry, Grayish Groundcherry

Physalis cinerascens
(M. Dunal)
A. S. Hitchcock var.
cinerascens

Growth Habit: Perennial.

Stems: Low growing, covered with branched and stellate hairs.

Leaves: Simple, alternate; blades ovate to round, margins entire or remotely sinuate; petioles densely pubescent with branched and stellate hairs.

Inflorescence: Flowers solitary in the leaf axils; pedicels pubescent; flowers pointed downward.

> **Calyx:** Sepals 5, united, densely pubescent, inflated and surrounding the fruit at maturity.

> **Corolla:** Petals 5, united, yellow on the outside, yellow and purple inside, pubescent within, actinomorphic.

> **Stamens:** 5, epipetalous, anthers whitish-purple.

> **Pistil:** Ovary superior; style 1, unbranched.

Fruit: A berry enclosed by an inflated calyx.

General Comments: Native. This species occurs in deep, sandy soils. The fruits, enclosed by an inflated calyx, resemble miniature husk tomatoes or tomatillos.

American Black Nightshade

Solanum americanum
P. Miller

Growth Habit: Annual; from a taproot.

Stems: Erect, branching freely, pubescent and with short, winglike ribs.

Leaves: Simple, alternate; blades ovate or ovate-lanceolate, pubescent, margins sinuate; petioles ribbed, pubescent.

Inflorescence: Flowers in axillary cymes along the stem axis.

    Calyx: Sepals 5, united, pubescent.

    Corolla: Petals 5, united, white or occasionally bluish tinged.

    Stamens: 5, epipetalous; anthers poricidal, yellow.

    Pistil: Ovary superior; style longer than the stamens.

Fruit: A black or dark purple berry.

General Comments: Native. American black nightshade is poisonous (Sperry et al. 1968). It occurs in a variety of disturbed sites but is more common on the margins of wet sites.

Pathogens and Pests: Artificial inoculations of *S. americanum* with *Xylella fastidiosa* (the causal agent of citrus variegated chlorosis) were 31 percent successful, suggesting that it is a possible weedy host for this phytoplasma (Lopes et al. 2002). CVC-causing *X. fastidiosa* is transmitted by the grass leafhopper (*Ferrariana trivittata*).

Redberry Nightshade

Solanum campechiense
C. Linnaeus

Growth Habit: Annual; from a taproot.

Stems: Usually low growing and branched; pubescent and spiny.

Leaves: Simple, alternate; blades oblong, lobed and spiny.

Inflorescence: A few-flowered cyme.

    Calyx: Sepals 5, united, spiny.

    Corolla: Petals 5, united, light purple.

    Stamens: 5, epipetalous.

    Pistil: Ovary superior.

Fruit: A mottled, greenish berry.

General Comments: Native. Redberry nightshade is painful to the touch. We have noted only greenish berries in this species. It occurs in heavy clays on the margins of freshwater and brackish marshes, primarily in Cameron County.

Silverleaf Nightshade, Trompillo

Solanum elaeagnifolium
A. Cavanilles

Growth Habit: Perennial.

Stems: Erect with stout prickles, green or gray with stellate pubescence.

Leaves: Simple, alternate; blades elliptic or lanceolate, grayish-green, stellate pubescent, margins entire; petioles pubescent, stipules present.

Inflorescence: A cyme usually near the apex of the stem axis.

> **Calyx:** Sepals 5, united, stellate pubescent and bearing numerous prickles.

> **Corolla:** Petals 5, united, purple or occasionally white with yellow, linear nectar guides below.

> **Stamens:** 5, epipetalous; anthers poricidal, yellow.

> **Pistil:** Ovary superior, pubescent, style longer than the stamens.

Fruit: A berry ripening from yellow-orange to black at maturity.

General Comments: Native. This is a widespread species in the drier areas of the United States and Mexico. Silverleaf nightshade forms colonies in arid, disturbed sites and contains the toxic alkaloid solanine (Sperry et al. 1968; Hart et al. 2003).

Pathogens and Pests: Experimental inoculations of *S. elaeagnifolium* with pepper mottle virus (PepMoV) have indicated that it is a host for this disease-causing agent of commercial chile peppers (Rodriguez-Alvarado et al. 2002). However, the mode of transmission is not known.

Buffalobur, Kansas Thistle, Spanish Thistle

Solanum rostratum
M. Dunal

Growth Habit: Annual; from a taproot.

Stems: Sprawling, all parts of plants with formidable, long, yellow, stout spines, some with purple bases, stellate pubescent.

Leaves: Simple, alternate; blades pinnatisect, spiny and stellate pubescence.

Inflorescence: A raceme; pedicels with stout spines and stellate pubescent.

    Calyx: Sepals 5, united with lanceolate lobes, spiny and pubescent.

    Corolla: Petals 5, united, actinomorphic, golden-yellow.

    Stamens: 5, epipetalous; filaments short; anthers poricidal.

    Pistil: Ovary superior; style 1, unbranched.

Fruit: A spiny, pubescent capsule.

General Comments: Native. Buffaloburs are more common in the arid regions of South Texas. The plant is poisonous to livestock, but they usually avoid this species (Sperry et al. 1968; Hart et al. 2003).

Texas Nightshade, Hierba Mora

Solanum triquetrum
A. Cavanilles

Growth Habit: Perennial.

Stems: Erect and often sprawling or supported by other plants, glabrous.

Leaves: Simple, alternate; blades lanceolate, acuminate at the apex, cordate or rounded at the base, glabrous, margins entire; petioles glabrous.

Inflorescence: Few flowered in a subterminal cyme.

>**Calyx:** Sepals 5, united, glabrous, often purple tinged.

>**Corolla:** Petals 5, united, white or bluish-white.

>**Stamens:** 5, epipetalous; anthers poricidal, yellow.

>**Pistil:** Ovary superior; style 1.

Fruit: A bright, red berry.

General Comments: Native. Texas nightshade is common in shaded sites. It often sprawls over taller shrubs.

URTICACEAE

Pennsylvania Pellitory, Blunt Pellitory, Hammerwort

Parietaria pensylvanica
G. H. Muhlenberg *ex*
C. von Willdenow

Growth Habit: Annual; shallowly rooted; plants monoecious.

Stems: Erect, pubescent with bent hairs.

Leaves: Simple, alternate; blades ovate, glabrous but pubescent on the entire margins; petioles pubescent.

Inflorescence: Flowers unisexual or occasionally bisexual and in axillary clusters.

> **Calyx:** Subtended by lanceolate, pubescent bracts; sepals 5, connate, pubescent.
>
> **Corolla:** Absent.
>
> **Stamens:** 3 (if present); anthers white; filaments whitish-translucent with cartilaginous striations.
>
> **Pistil:** Ovary superior.

Fruit: An achene.

General Comments: Native. Pennsylvania pellitory occurs in moist, shaded sites during the cool season.

Heartleaf Stingingnettle, Ortiguilla

Urtica chamaedryoides
F. Pursh

Growth Habit: Annual; from a taproot; monoecious or dioecious; all parts with stinging, capillary hairs.

Stems: Erect, ridged, reddish at maturity with translucent, spinelike hairs enlarged at the base and tapered into a sharp spine.

Leaves: Simple, opposite; blades ovate to ovate-lanceolate, spiny, margins serrate; petioles present.

Inflorescence: Flowers in globular heads.

> **Calyx:** Sepals 4, free, with stinging hairs.
>
> **Corolla:** Petals absent.
>
> **Staminate Flowers:** Stamens 4; pollen white.
>
> **Pistillate Flowers:** Ovary superior.

Fruit: An achene.

General Comments: Native. This species is found during the cool season. The spiny, glasslike hairs release histamine and cause painful skin irritations that often last several days after contact.

VERBENACEAE

Gulf Coast Mock Vervain, Beaked Mock Vervain, Pale Mock Vervain

Glandularia quadrangulata (A. A. Heller) R. Umber

Syn. *Verbena quadrangulata* A. A. Heller

Growth Habit: Annual; from a taproot.

Stems: Prostrate, ribbed or angled, densely pubescent.

Leaves: Simple, opposite; blades deeply lobed, pubescent; petioles pubescent.

Inflorescence: Flowers subtended by a pubescent bract and arranged in a loosely flowered spike below but in a crowded spike above.

> **Calyx:** Sepals 5, united, pubescent.

> **Corolla:** Petals 5, united below and lobed above, bluish-white, pubescent in the throat.

> **Stamens:** 4, epipetalous.

> **Pistil:** Ovary superior, 4 lobed.

Fruit: Separating into 4 bony, dumbbell-shaped nutlets.

General Comments: Native. This low-growing species occurs on field margins and roadsides during the cool season.

Small Blue Vervain, Texas Vervain

Verbena halei J. K. Small

Syn. *V. officinale*
C. Linnaeus subsp. *halei*
(J. K. Small) S. Barber

Growth Habit: Annual or perennial.

Stems: Erect, ribbed, not round, appressed pubescent.

Leaves: Simple, opposite; blades oblanceolate below and lobed, linear and lobed above, pubescent; petioles winged.

Inflorescence: Flowers subtended by a pubescent bract and with a loosely arranged spike.

> **Calyx:** Sepals 5, united, appressed pubescent, purple tinged above.
>
> **Corolla:** Petals 5, united into a tube, blue, pubescent in the throat, regularly symmetrical or nearly so.
>
> **Stamens:** 4, epipetalous, minute.
>
> **Pistil:** Ovary superior, 4 lobed.

Fruit: Separated into 4 bony, cylindrical nutlets.

General Comments: Native. Small blue vervain is a common Texas weed. It is most common during the spring but is found sporadically during the warm season.

Fanleaf Vervain

Verbena plicata
E. Greene

Growth Habit: Perennial.

Stems: Erect, densely pubescent with glandular and nonglandular hairs.

Leaves: Simple, opposite; blades spatulate, veiny, pubescent, margins lobed and toothed and gradually tapered to a petiole.

Inflorescence: A spike with each flower subtended by a coarsely pubescent bract.

    Calyx: Sepals 5, united, glandular, and pubescent.

    Corolla: Petals 5, slightly zygomorphic, dark blue, pubescent.

    Stamens: 4, epipetalous.

    Pistil: Ovary superior; style 1.

Fruit: 4 lobed and nutlike.

General Comments: Native. Fanleaf vervain is common on the margins of dry roadsides.

VITACEAE

Ivy Treebine, Possum Grape, Marine Ivy

Cissus incisa
C. Des Moulins

Growth Habit: Perennial; a succulent, malodorous vine, woody below.

Stems: Twining with apical tendrils, glabrous, succulent.

Leaves: Trifoliolate, alternate but usually with tendrils on the opposite surface of the stems; leaflets succulent, lobed and toothed; petioles grooved; stipules present.

Inflorescence: Flowers in corymbose or umbellate cymes.

    Calyx: Sepals 4, united, minute.

    Corolla: Petals 4, free, actinomorphic, yellow-green.

    Stamens: 4, free; anthers yellow.

    Pistil: Ovary superior; style 1, unbranched.

Fruit: A purple or black, fleshy berry that resembles a small grape.

General Comments: Native. Ivy treebine is a foul-smelling vine on fences in a variety of disturbed sites.

Waterwithe Treebine, Bejuco Loco

Cissus sicyoides
C. Linnaeus

Growth Habit: Perennial; an elongated woody vine.

Stems: Young stems dark green, glabrous, older stems with reddish-purple papillae and cream-colored lenticels, bark peeling and exposing green stems, nodes swollen.

Leaves: Simple, alternate, succulent, with tendrils opposite some leaves; blades ovate to broadly rounded, glabrous, margins undulate; petioles glabrous, elongated.

Inflorescence: Cymose from leaf axils.

> **Calyx:** Cuplike, green, reduced to a rim around the ovary.

> **Corolla:** Petals 4, yellow-green, connate (not spreading), attached to a floral disk.

> **Stamens:** 4; anthers white.

> **Pistil:** Surrounded by a nectar-bearing disk; ovary superior; style 1, unbranched.

Fruit: A black or purple berry with only 1 seed.

General Comments: Introduced. This species was found climbing on citrus trees in the Weslaco area (Hidalgo County) during the fall of 2003. It represents a possible problem for citrus production if it becomes widely established in the LRGV (French, Lonard, and Everitt 2003).

ZYGOPHYLLACEAE

California Caltrop

Kallstroemia californica
(S. Watson) A. Vail

Growth Habit: Annual; from a taproot.

Stems: Prostrate, pubescent with 2 types of hairs, the shorter more numerous and curly, the longer straight and perpendicular or angled from the stem.

Leaves: Even-pinnately compound, opposite, slightly unequal leaf sizes; leaflets green above and silver-gray on the lower epidermis, pubescent; stipules linear, pubescent.

Inflorescence: Flowers solitary from the leaf axils; pedicels pubescent.

> **Calyx:** Sepals 5, free, densely pubescent.
>
> **Corolla:** Petals 5, free, yellow-orange.
>
> **Stamens:** 10, free; anthers yellow.
>
> **Pistil:** Ovary superior, knobby, not spiny.

Fruit: A pubescent, knobby capsule with a persistent style.

General Comments: Native. This low-growing weed occurs during the warm season. It is similar to *Tribulus terrestris* but lacks spiny fruits. Sperry et al. (1968) indicate that it is poisonous to rabbits, goats, and cattle. Hart et al. (2003) report that wilted plants are more toxic to livestock than well-watered plants.

Puncturevine, Goathead, Abrojo De Flor Amarilla

Tribulus terrestris
C. Linnaeus

Growth Habit: Annual; from a taproot.

Stems: Prostrate, pubescent with 2 types of hairs, the shorter more numerous, bent, the longer straight and nearly perpendicular to the stem.

Leaves: Even-pinnately compound, opposite with unequal leaf sizes; leaflets pubescent on the midvein and margins above, densely pubescent on the lower epidermis; stipules linear, pubescent.

Inflorescence: Flowers usually solitary from the leaf axils; pedicels pubescent.

> **Calyx:** Sepals 5, free, pubescent.
>
> **Corolla:** Petals 5, free, yellow.
>
> **Stamens:** 10, free; anthers yellow, pubescent.
>
> **Pistil:** Ovary superior, style branched into 5 stigmatic areas (resembles a starfish).

Fruit: A spiny capsule.

General Comments: Introduced. This species is common on roadsides; in agricultural fields; and in other dry, disturbed sites. Translated from Latin, *Tribulus terrestris* means "trouble on the ground." The spiny fruits readily adhere to clothing, shoes, and tires. Occasionally, fruit fragments will be found as foreign material in pinto bean packages. Vegetative portions of the plant are toxic to livestock, especially sheep (Sperry et al. 1968). It is controlled in California and Texas by stem and seed weevils (*Microlarinus lypriformis* and *M. lareynii*, respectively) (Goeden and Kirkland 1978; Kirkland and Goeden 1978; Rummel and Arnold 1992). Propagules may be buried for several years and remain viable.

Class Liliiopsida

MONOCOTS

ALLIACEAE

Crow Poison, False Garlic

Nothoscordum bivalve
(C. Linnaeus) N. Britton

Growth Habit: Perennial; from a scaly bulb.

Stems: Reduced to a scape; aerial shoots from a subterranean bulb.

Leaves: Simple, basal; blades, linear, glabrous, succulent.

Inflorescence: Flowers in an umbellate cluster from a glabrous, elongated scape; a scaly bract present at the base of the scape.

> **Tepals:** Sepals and petals similar in size and color; 6 free, white, often with a purple longitudinal midvein.

> **Stamens:** 6; anthers yellow.

> **Pistil:** Ovary superior; style 1, unbranched.

Fruit: A 3-lobed capsule.

General Comments: Native. Although this species is called crow poison, there are no documented cases of human or animal deaths. It is a cool-season weed that occurs on roadsides and in lawns.

ARACEAE

Waterlettuce, Waterbonnet

Pistia stratiotes
C. Linnaeus

Growth Habit: Perennial; free floating or occasionally rooted; plants monoecious.

Stems: Aerial stems absent; rhizomes present.

Leaves: Simple, in a rosette; blades spongy, obovate, softly pubescent, light yellowish-green.

Inflorescence: Not observed.

> **Calyx:** Absent.
>
> **Corolla:** Absent.
>
> **Stamens:** Usually 6.
>
> **Pistil:** Ovary superior.

Fruit: Fruiting has not been reported in South Texas.

General Comments: Introduced. Waterlettuce is one of the most cosmopolitan weeds in the world. It is found on every continent except Europe and Antarctica. It is believed to be native to South America (Cordo, Deloach, and Ferner 1981). This species is a common aquatic weed in southeastern Texas and is present in the Rio Grande near Brownsville, Texas (Cameron County), and in Lake Corpus Christi near Mathis, Texas (San Patricio County). It forms large mats that cause waterways to become clogged and access to fishing, swimming, and boating to be reduced or eliminated (Stoddard 1989). Waterlettuce weevils (*Neohydronomous affinia*) provide effective control of waterlettuce (Deloach, Deloach, and Cordo 1976).

COMMELINACEAE

Upright Dayflower, Whitemouth Dayflower, Hierba Del Pollo

Commelina erecta
C. Linnaeus var. *erecta*

Syn. *C. elegans* K. Kunth

Growth Habit: Perennial.

Stems: Erect and sprawling, glabrous, reddish, rooting at the nodes.

Leaves: Simple, alternate; blades broadly lanceolate, glabrous, parallel veined, margins entire; petioles short; sheaths pubescent on the margins near the apex.

Inflorescence: Flowers in terminal clusters subtended by a minutely pubescent, folded, spathelike bract.

    Calyx: Sepals 3, translucent, free.

    Corolla: Petals 3, unequal (2 large and 1 smaller), free, fragile.

    Stamens: 3.

    Pistil: Ovary superior; style 1.

Fruit: A 3-lobed capsule.

General Comments: Native. This species is common in flower beds and is found occasionally on roadsides. It is difficult to eradicate.

CYPERACEAE

Purple Nutsedge

Cyperus rotundus C. Linnaeus

Growth Habit: Perennial; with an extensive rhizome system with irregularly spaced, tuberous (nutlike) extensions.

Stems: Erect, triangular, glabrous.

Leaves: Simple, 3 ranked, flat, glabrous; sheaths closed, ligules absent.

Inflorescence: Wind-pollinated flowers arranged in reddish spikelets.

> **Glume (Bract):** 1, reddish, but with a green keel subtending each flower.

> **Calyx:** Absent.

> **Corolla:** Absent.

> **Pistil:** Ovary superior; style branches 3.

Fruit: An achene.

General Comments: Introduced. Purple nutsedge is one of the world's worst weeds. It is common in a wide variety of disturbed sites, including agricultural fields, flower beds, and cracks and crevices of sidewalks.

HYDROCHARITACEAE

Hydrilla, Waterthyme, Whorled Waterthyme

Hydrilla verticillata (C. von Linne) J. Royle

Growth Habit: Perennial; submersed, rhizomatous aquatic rooted to the bottom, but fragments break loose and are free floating; plants monoecious or dioecious.

Stems: Profusely branching near the water surface and becoming stoloniferous.

Leaves: Simple, whorled; blades narrowly lanceolate, midvein on the lower epidermis spiny or glandular.

Inflorescence: Flowers borne singly from a spathe (flowering is seldom observed in the LRGV).

> **Staminate Flowers:**
>
> > **Calyx:** Sepals 3, reddish-white or brownish.
> >
> > **Corolla:** Petals 2, reddish-white.
> >
> > **Stamens:** 3.
>
> **Pistillate Flowers:**
>
> > **Calyx:** Sepals 3, white.
> >
> > **Corolla:** Petals 3, translucent.

Fruit: Usually does not form fruits or seeds; it forms compact dormant buds called turions in the leaf axils, and subterranean tubers on rhizomes.

General Comments: Introduced. Hydrilla is a federally listed noxious weed that was initially documented in Florida in 1960 (Langeland 1996). It was introduced to sell in pet stores for aquariums. Growth rates of 2.54 cm per day have been reported. It is an aggressive invader in the Rio Grande, Arroyo Colorado, and canals. It can clog bodies of water at depths of 5.0 to 6.5 m in one year.

Biocontrol: Both *Plectosporium tabacinum* (*Plectosphaerella cucumerina*) and *Fusarium culmorum* have been exploited as fungal biological control agents for *H. verticillata* (Smither-Kopperl, Charudattan, and Berger 1998a, 1998b). However, these control measures have been largely ineffective due to the rapid stoloniferous spread of the plant. Triploid grass carp provide excellent control of hydrilla (Anonymous 2001).

POACEAE[1]

Bushy Bluestem, Clustered Bluestem

Andropogon glomeratus (T. Walter) N. Britton, E. Sterns, & J. Poggenburg

Growth Habit: Perennial.

Culms: Erect, up to 1.3 m tall, mostly glabrous but with a few scattered hairs, flat in cross section.

Leaves:

Sheaths: Keeled, glabrous but with a few scattered pustular glands.

Ligules: Membranous with a ring of hairs at the apex.

Blades: Flat, yellowish-green, glabrous.

Inflorescence: Racemose with branches arising from sheathing bracts.

Spikelets: Paired; fertile spikelet sessile; reduced spikelets reduced to a pilose pedicel.

Glumes: As long as the spikelet, awnless.

Lemmas: Awns 1, bent.

Paleas: Membranous, inconspicuous.

Caryopses: Minute.

General Comments: Native. This species is conspicuous in the fall and early winter in roadside ditches and in moist depressions. It is common in southeastern Texas and is occasionally seen as a lawn ornamental in the Houston area.

[1]Treatment of grasses is modified from Lonard (1993).

Sixweeks Threeawn, Ascending Threeawn

Aristida adscensionis
C. Linnaeus

Growth Habit: Annual.

Culms: Geniculate or erect, branching above the base.

Leaves:

> **Sheaths:** Closed, glabrous.
>
> **Ligules:** A pubescent ring.
>
> **Blades:** Involute.

Inflorescence: Linear, branches appressed or ascending, several per node.

Spikelets: In dense clusters, florets 1, disarticulation above the glumes, pedicels glabrous.

> **Glumes:** Unequal, glabrous.
>
> **Lemmas:** Awns 3, up to 1.5 cm long.
>
> **Paleas:** Inconspicuous.
>
> **Caryopses:** Linear, as long as the lemma.

General Comments: Native. This species is often found on dry, caliche sites. An abundance of species of *Aristida* in rangelands is often an indication of heavy grazing pressure. These species have low nutritional values for grazing animals.

Purple Threeawn

Aristida purpurea T. Nuttall var. *purpurea*

Syn. *A. roemeriana* G. Scheele

Growth Habit: Perennial; with tough, knotty bases.

Culms: Erect or geniculate, unbranched above the base.

Leaves:

> **Sheaths:** Open and glabrous.
>
> **Ligules:** A minute, fringed membrane.
>
> **Blades:** Involute, minutely pubescent above and with a few scattered hairs on the lower epidermis.

Inflorescence: A linear panicle with appressed or ascending branches.

Spikelets: Florets 1, disarticulation above the glumes, pedicels glabrous.

> **Glumes:** Unequal, linear, glabrous, awnless.
>
> **Lemmas:** As long as the spikelet, purple, glabrous, 3 awned, the awns up to 2.5 cm long.
>
> **Paleas:** Inconspicuous.
>
> **Caryopses:** Linear.

General Comments: Native. Two additional varieties of *A. purpurea* are present in the LRGV: *A. purpurea* T. Nuttall var. *longiseta* (E. von Steudel) G. Vasey (long-awned purple threeawn) and *A. purpurea* T. Nuttall var. *wrightii* (G. Nash) K. Allred (Wright's purple threeawn). Variety *longiseta* has awns 6.0–7.5 cm long, and variety *wrightii* has awns 1.0–1.5 cm long. All varieties are common on dry sites and are often abundant on sandy roadsides.

Giant Reed, Carrizo

Arundo donax C. Linnaeus

Growth Habit: Perennial; with thick rhizomes.

Culms: Erect, 3.5–4.0 m tall, glabrous, hollow at the internodes.

Leaves:

> **Sheaths:** Open and scabrous on the margins.
>
> **Ligules:** A glabrous, fringed membrane.
>
> **Blades:** Elongated, flat, green, glaucous, glabrous.

Inflorescence: A large panicle.

Spikelets: Laterally compressed, florets 2–4, disarticulation above the glumes and between the florets, pedicels glabrous; rachillas glabrous.

> **Glumes:** Subequal, lanceolate, glabrous, awnless.
>
> **Lemmas:** Awns 1, silky pilose.
>
> **Paleas:** Membranous, pubescent.
>
> **Caryopses:** Absent.

General Comments: Introduced. This species was initially introduced by the Texas State Highway Department for erosion control. Flowering occurs in late summer and fall. The species has spread from roadsides to riverbanks and canal banks. Giant reed is a severe threat to riparian areas, where it displaces native plants and animals by forming massive stands. It also alters channel morphology by retaining sediment and constricting flows and may reduce navigability (Dudley 2000). It consumes excessive amounts of water, and growth rates of 5 cm per day have been reported (Perdue 1958; Bell 1997). This species is invasive in the riparian zone of the Rio Grande and is a management problem in the Del Rio, Texas, area. Giant reed is similar to *Phragmites australis* but has glabrous spikelet rachillas. It forms monotypic stands and crowds out native species and increases the potential for wildfires. Glyphosate applied broadcast or as a basal treatment to cut culms has been the most successful control method (Dudley 2000).

Pathogens and Pests: Several insect pests and diseases that are commonly encountered on sugarcane, maize, and sorghum (and other cereals) are also seen on *A. donax*. See Appendix 3 for a more complete listing.

King Ranch Bluestem

Bothriochloa ischaemum
(C. Linnaeus) Y. Keng
var. *songarica*
(F. von Fischer &
C. von Meyer)
R. Celarier & J. Harlan

Growth Habit: Perennial.

Culms: Geniculate, branching from the base, rooting at the nodes, glabrous.

Leaves:

> **Sheaths:** Overlapping.
>
> **Ligules:** Glabrous, about 1 mm long.
>
> **Blades:** Flat or folded, with scattered hairs above and glabrous on the lower epidermis.

Inflorescence: A panicle with digitate, racemose branches; rachis with pilose hairs below the sterile spikelet.

Spikelets: Sterile spikelet slightly shorter than the fertile spikelet, pedicels pilose.

> **Glumes:** The first pubescent, the second glabrous.
>
> **Lemmas:** As long as the floret, membranous, with a twisted and bent awn arising from the base.
>
> **Paleas:** Inconspicuous.
>
> **Caryopses:** Minute.

General Comments: Introduced. King Ranch bluestem is more common in the northern portions of South Texas, where it dominates roadsides and pastures.

Red Grama

Bouteloua trifida
G. Thurber

Growth Habit: Perennial; tufted.

Culms: Low growing, wiry, erect or geniculate, unbranched.

Leaves:

> **Sheaths:** About 2/3 length of the internodes.
>
> **Ligules:** A minute ring of hairs.
>
> **Blades:** Flat or usually involute, scabrous above and glabrous on the lower epidermis, margins with a few pilose hairs.

Inflorescence: With persistent racemose branches; branches spicate with spikelets on lower surface of rachis, branches 1 per node, rachis appressed pubescent.

Spikelets: Florets 2, pedicels scabrous, rachilla glabrous.

> **Glumes:** Glabrous, purple, awnless.
>
> **Lemmas:** Glabrous above and silky pubescent below, awns 3.
>
> **Paleas:** About as long as the lemmas.
>
> **Caryopses:** Linear, yellowish-brown.

General Comments: Native. Red grama is common on well-drained soils in the drier areas of South Texas. It is tolerant of drought conditions.

Rescuegrass

Bromus catharticus
M. A. Vahl

Syn. *B. unioloides*
(C. von Willdenow)
K. Kunth; *B. willdenowii*
K. Kunth

Growth Habit: Annual.

Culms: Erect or geniculate, hollow at the internodes, unbranched, glabrous.

Leaves:

> **Sheaths:** Open, glabrous below and pilose above.
>
> **Ligules:** Membranous, glabrous.
>
> **Blades:** Flat, nearly glabrous above and sparsely pubescent on the lower epidermis.

Inflorescence: An open or contracted panicle, branches 1–2 per node; spikelets borne near the tips of the branches.

Spikelets: Laterally compressed, florets 8–12, disarticulation above the glumes and between the florets; pedicel and rachilla scabrous.

> **Glumes:** Glabrous, awnless.
>
> **Lemmas:** Glabrous, awn tipped.
>
> **Paleas:** Slightly shorter than the lemma, acute at the apex.
>
> **Caryopses:** Lanceolate, golden-brown.

General Comments: Introduced. Rescuegrass is a common cool-season species that occurs in lawns and along roadsides.

Pathogens and Pests: *Bromus catharticus* is susceptible to a common cereal powdery mildew fungus (*Blumeria* [*Erysiphe*] *graminis*). This fungus is an obligate parasite of the plant. Therefore, it cannot be cultured on artificial media in the laboratory. A whitish-gray mycelium appears on the upper leaf epidermis of the leaves, and occasionally sexual fruiting structures of the fungus are embedded in the hyphae. In susceptible grasses, the fungus causes most of the plant to become chlorotic. Death of the plant may ensue, but only in extreme cases.

Southern Sandbur

Cenchrus echinatus
C. Linnaeus

Growth Habit: Annual.

Culms: Geniculate, branching above the base, rooting at the lower nodes.

Leaves:

> **Sheaths:** Glabrous, about 3/4 the length of the internodes.
>
> **Ligules:** A short ring of hairs.
>
> **Blades:** Flat, glabrous, margins scabrous.

Inflorescence: Spicate, burlike clusters.

Spikelets: 4 or more enclosed in a tough bur, spines of the burs flattened near the base, retrorsely barbed.

> **Glumes:** Whitish, about the same length as the spikelet, membranous, glabrous, awnless.
>
> **Lemmas:** Membranous, glabrous.
>
> **Paleas:** Membranous.
>
> **Caryopses:** Usually not seen unless the burs are dissected.

General Comments: Native. Southern sandburs are common in sandy loam sites and on the margins of cultivated fields. The burs detach easily and adhere to animal hair and clothing.

Coastal Sandbur, Common Sandbur, Southern Sandbur

Cenchrus spinifex
A. Cavanilles

Syn. *C. incertus*
M. A. Curtis;
C. pauciflorus
G. Bentham

Growth Habit: Annual; tufted.

Culms: Geniculate.

Leaves:

>**Sheaths:** Pubescent, margins pilose, about 1/2 the length of the internodes.

>**Ligules:** A ring of hairs.

>**Blades:** Folded or flat, glabrous or with pustulate-based hairs above and on the lower epidermis.

Inflorescence: Spikelets in burs and contracted on the axis, rachis angled.

Spikelets: 2–3 spikelets enclosed in tough burs, burs and bristles pubescent with recurved spines and bristles.

>**Glumes:** White, membranous, glabrous, awnless.

>**Lemmas:** Membranous and glabrous.

>**Paleas:** Membranous.

>**Caryopses:** Usually not seen.

General Comments: Native. Coastal sandbur is found in a wide variety of disturbed sites but is most abundant on sandy soils. It adheres to animal hair, clothing, and shoes. It is much more painful to the touch than *C. echinatus* and is one of the worst weeds in South Texas.

Slimspike Windmillgrass

Chloris andropogonoides
E. Fournier

Growth Habit: Perennial; tufted or stoloniferous.

Culms: Geniculate, branched near the base.

Leaves:

    Sheaths: 3/4 the length of the internodes.

    Ligules: A fringed membrane with some hairs at the apex.

    Blades: Folded, glabrous above and with scattered pustulate-based hairs on the lower epidermis.

Inflorescence: Disarticulating below and becoming a tumbleweed, digitate; rachis with a tuft of pilose hairs at the base; antrorsely scabrous, spikelets borne on the lower surface of the rachis.

Spikelets: Florets 2, the uppermost reduced and sterile.

    Glumes: Lanceolate, glabrous, awnless.

    Lemmas: Greenish-yellow, margins pilose, awn 1, bent above.

    Paleas: About as long as the lemmas, membranous.

    Caryopses: Lanceolate, minute.

General Comments: Native. This species is found in weedy lawns and on roadsides. At maturity the inflorescence breaks off and forms a tumbleweed.

Swollen Fingergrass

Chloris barbata
O. Swartz

Syn. *C. inflata* J. Link

Growth Habit: Annual; tufted.

Culms: Erect or geniculate, flattened near the base, branching above the base and occasionally rooting at the lower nodes.

Leaves:

    Sheaths: Less than 1/3 the length of the internodes.

    Ligules: A fringed, glabrous membrane.

    Blades: Flat, glabrous, margins serrate with pilose hairs near the base.

Spikelets: Oblanceolate, florets 3, disarticulation above the glumes and between the florets, pedicels pubescent and scabrous, a dense tuft of hairs below the fertile floret.

    Glumes: Translucent, linear-lanceolate, glabrous and awn tipped.

    Lemmas: Awned, purple tinged, margins densely pilose.

    Paleas: About as long as the lemmas, membranous, glabrous.

    Caryopses: Lanceolate, yellow.

General Comments: Native. This weedy grass is found only in the southern fringe of counties in South Texas. It is abundant in weedy lawns and in cracks and crevices of pavement. It grows rapidly after mowing.

Paraguay Windmillgrass

Chloris canterae J. Arechavleta

Growth Habit: Perennial; tufted.

Culms: Erect, unbranched, up to 1 m tall.

Leaves:

> **Sheaths:** About 1/2 the length of the internode, glabrous.
>
> **Ligules:** A ring of hairs.
>
> **Blades:** Flat or folded, glabrous, margins scabrous.

Inflorescence: Branches digitate with 2 or more per node, pubescent or glabrous, bearing spikelets to the base of each branch, spikelets borne on the lower surface of the rachis.

Spikelets: Ovate, pedicels scabrous, florets 3, the uppermost sterile.

> **Glumes:** Lanceolate, dark brown.
>
> **Lemmas:** Ovate-lanceolate, pilose on the midvein and margins.
>
> **Paleas:** About 3/4 the length of the lemmas, margins pilose.
>
> **Caryopses:** Ovate or 3 angled.

General Comments: Introduced. Paraguay windmillgrass is a roadside weed.

Fringed Windmillgrass

Chloris ciliata O. Swartz

Growth Habit: Perennial; tufted.

Culms: Erect or geniculate, branching above the base and occasionally rooting at the lower nodes.

Leaves:

> **Sheaths:** About 3/4 the length of the internodes, glabrous.

> **Ligules:** A fringed, glabrous membrane.

> **Blades:** Flat, glabrous, margins scabrous.

Inflorescence: Branches digitate, spikelets borne on the lower surface of the rachis.

Spikelets: Oblanceolate, florets 2, the uppermost sterile, pedicels glabrous.

> **Glumes:** Lanceolate, brown, glabrous, awnless.

> **Lemmas:** Lemma of the fertile floret pubescent, awned; lemma of the sterile floret awnless.

> **Paleas:** About as long as the lemma, margins pubescent.

> **Caryopses:** Lanceolate-elliptic, amber.

General Comments: Native. This species is present on roadsides and in other disturbed sites.

Nash's Windmillgrass

Chloris × subdolichostachya J. K. A. Müller (*cucullata × verticillata*)

Syn. *C. latisquamea* G. Nash

Growth Habit: Perennial; tufted or stoloniferous.

Culms: Erect, flattened at the base.

Leaves:

> **Sheaths:** Almost as long as the internodes.
>
> **Ligules:** A fringed membrane with minute hairs at the apex.
>
> **Blades:** Flat or folded, grayish-green, glabrous; margins scabrous.

Spikelets: Lanceolate, florets 2, pedicels scabrous.

> **Glumes:** Lanceolate, purple tinged, awnless.
>
> **Lemmas:** Awned, purple tinged; margins pubescent.
>
> **Paleas:** About as long as the lemmas, membranous.
>
> **Caryopses:** Minute, fertile.

General Comments: Native. Nash's windmillgrass is found in a variety of disturbed sites. It is a naturally occurring, fertile hybrid of *C. cucullata* and *C. verticillata*.

Bermudagrass

Cynodon dactylon
(C. Linnaeus)
C. Persoon

Growth Habit: Perennial; stoloniferous and rhizomatous.

Culms: Prostrate to erect, rooting at the nodes.

Leaves:

> **Sheaths:** Keeled.
>
> **Ligules:** A ring of hairs, pilose on the margins.
>
> **Blades:** Flat or folded, margins minutely serrate with pilose hairs near the base.

Inflorescence: A panicle with digitate, unilateral, spicate branches.

Spikelets: Lanceolate, florets 1, pedicels subsessile.

> **Glumes:** Linear-lanceolate, glabrous, awnless.
>
> **Lemmas:** Awnless.
>
> **Paleas:** About the same length as the lemmas, glabrous.
>
> **Caryopses:** Minute.

General Comments: Introduced. Bermudagrass and St. Augustine grass are the only two practical choices for turf grasses in South Texas. However, bermudagrass is an aggressive weed in a variety of disturbed sites. Bermudagrass may be toxic to cattle in the fall if it is growing on moldy plant debris (Hart et al. 2003). It is the most widely distributed plant in the world.

Pathogens and Pests: *Cynodon dactylon* is infected with a smut fungus (*Ustilago cynodontis*) that usually appears in mid- to late spring. Smutted inflorescences are covered by black spores of the fungus, giving the inflorescence a dark paintbrush-like appearance. The peridium, or outer layer, of the smut gall rarely remains intact for more than a few days, which gives the smutted inflorescence this appearance. The smut fungus grows systemically from infected caryopses. See Appendix 4 for further information.

Crowfootgrass

Dactyloctenium aegyptium (C. Linnaeus) A. Palisot de Beauvois

Growth Habit: Annual; occasionally rooting at the lower nodes.

Culms: Erect or spreading, unbranched, glabrous.

Leaves:

> **Sheaths:** As long as the internode, glabrous.

> **Ligules:** A membrane fringed with hairs.

> **Blades:** Flat, with papillose hairs above and below, margins slightly crisped with papillose hairs.

Inflorescence: A panicle with digitately arranged, 1-sided spicate branches, rachis extended as a sharp, naked point beyond the spikelets.

Spikelets: Laterally compressed, florets 3–5, sterile florets above.

> **Glumes:** About equal, glabrous, the second awned.

> **Lemmas:** Awned, glabrous but scabrous on the veins.

> **Paleas:** About as long as the lemmas, membranous.

> **Caryopses:** Round, rugose, cream colored.

General Comments: Introduced. This species is common in lawns and flower beds during the warm season and into early winter.

Kleberg Bluestem

Dichanthium annulatum
(P. Forsskål) O. Stapf

Growth Habit: Perennial; stoloniferous.

Culms: Prostrate and becoming erect when flowering, branching from the base, with a ring of densely tufted hairs at the nodes.

Leaves:

> **Sheaths:** Less than 1/2 the length of the internodes.

> **Ligules:** Membranous and glabrous.

> **Blades:** Flat, margins glabrous, pilose at the base.

Inflorescence: A panicle with digitate, racemose branches, branches 1 or 2 per node; peduncle glabrous or nearly so.

Spikelets: Paired with 1 sessile, awnless and fertile and 1 pedicelled, awned and sterile.

> **Glumes:** Purple tinged, as long as the spikelet.

> **Lemmas:** Awned, translucent.

> **Paleas:** As long as the lemmas, membranous.

> **Caryopses:** Ovoid, brown.

General Comments: Introduced. Kleberg bluestem is an invasive, drought- and fire-tolerant species. It occurs in fallow fields, in vacant lots, and on roadsides. It often forms monotypic stands on roadsides during the fall and early winter. Kleberg bluestem is similar to *D. aristatum,* but the peduncle below the spikelets is glabrous or nearly so.

Angleton Bluestem

Dichanthium aristatum
(J. Poiret) C. Hubbard

Growth Habit: Perennial; stoloniferous.

Culms: Erect or geniculate, branching, glabrous but densely pubescent at the nodes.

Leaves:

> **Sheaths:** About 1/2 the length of the internodes.
>
> **Ligules:** Membranous and glabrous.
>
> **Blades:** Flat, margins glabrous, pilose at the base.

Inflorescence: A panicle with digitate, racemose branches, branches 1 or 2 per node; peduncle pubescent.

Spikelets: Paired with 1 sessile, awnless and fertile and 1 pedicelled, awned and sterile.

> **Glumes:** Purple tinged, as long as the spikelet.
>
> **Lemmas:** Awned, translucent.
>
> **Paleas:** As long as the lemmas, membranous.
>
> **Caryopses:** Ovoid and brown.

General Comments: Introduced. Angleton bluestem is similar to *D. annulatum,* but the peduncle below the spikelets is silky pubescent.

Asian Crabgrass, Twohorned Crabgrass

Digitaria bicornis (J. de Lamarck)
J. J. Römer & J. A. Schultes

Growth Habit: Annual.

Culms: Geniculate, rooting at the lower nodes.

Leaves:

> **Sheaths:** 1/3 to 1/2 the length of the internodes, pubescent.
>
> **Ligules:** A fringed, glabrous membrane.
>
> **Blades:** Flat, margins minutely scabrous and with long hairs near the base.

Inflorescence: A panicle with 1-sided racemose branches, digitate.

Spikelets: Laterally compressed, paired on the rachis; pedicels serrate.

> **Glumes:** Awnless, margins silky pubescent.
>
> **Lemmas:** Glabrous, lemma of the fertile floret awn tipped.
>
> **Paleas:** As long as the lemma.
>
> **Caryopses:** Broadly ovate, yellowish-brown.

General Comments: Introduced. Asian crabgrass occurs in a variety of disturbed sites, including lawns, flower beds, and agricultural fields. It is similar to *D. ciliaris*.

Fall Witchgrass

Digitaria cognata
(J. A. Schultes) R. Pilger

Syn. *Leptoloma
cognatum*
(J. A. Schultes)
M. A. Chase

Growth Habit: Perennial; tufted with hard, knotlike bases.

Culms: Geniculate, branched near the base.

Leaves:

> **Sheaths:** Open, glabrous.
>
> **Ligules:** Membranous, glabrous.
>
> **Blades:** Flat, glabrous, often purple tinged.

Inflorescence: A broad, open panicle, disarticulating at the base and becoming a tumbleweed.

Spikelets: Widely spaced on the panicle branches, dorsally compressed (round in cross section), florets 2, disarticulation below the glumes.

> **Glumes:** The first minute, the second as long as the spikelet, glabrous, but pubescent on the margins, awnless.
>
> **Lemmas:** Lemma of the fertile floret, dark brown, indurate, lanceolate, margins enclosing the palea.
>
> **Paleas:** Marginally enclosed by the lemma.
>
> **Caryopses:** Ovate.

General Comments: Native. Fall witchgrass is usually found in dry sites. It is widespread as a weed in the temperate zone.

Junglerice

Echinochloa colonum
(C. Linnaeus) J. Link

Growth Habit: Annual; tufted, often rooting at the lower nodes.

Culms: Erect, branching above the base, glabrous but slightly pubescent at the nodes.

Leaves:

> **Sheaths:** Nearly as long as the internodes, margins with a few scattered hairs.
>
> **Ligules:** Absent.
>
> **Blades:** Flat, glabrous.

Inflorescence: A contracted, spicate panicle; peduncle with scattered, pilose hairs.

Spikelets: Dorsally compressed (round in cross section), ovate, pointed at the apex.

> **Glumes:** Ovate, finely pubescent, the second with pilose margins.
>
> **Lemmas:** Ovate with a short, mucronate tip.
>
> **Paleas:** About as long as the lemma, partially enclosed by the lemma.
>
> **Caryopses:** Inconspicuous.

General Comments: Introduced. Numerous authors refer to this species as *E. colona* (C. Linnaeus) J. Link. Junglerice is a warm-season species that occurs in flower beds, fields, and other disturbed sites. It is common in rice fields in southeastern Texas.

Pathogens and Pests: *Exserohilum monoceras* has been indicated as a fungal bioherbicide for control of *E. colonum* because it causes post-emergent damping-off and seedling death (Zhang and Watson 1997).

Goosegrass

Eleusine indica
(C. Linnaeus) J. Gaertner

Growth Habit: Annual; occasionally rooting at the nodes.

Culms: Erect or spreading, unbranched, glabrous.

Leaves:

> **Sheaths:** 3/4 the length of the internodes, nearly glabrous but with a few scattered hairs, margins with hairs near the base of the blades.
>
> **Ligules:** A fringed, glabrous membrane.
>
> **Blades:** Flat and glabrous.

Inflorescence: A panicle with digitately arranged, 1-sided, spicate branches, branches 1–4 per node, bearing spikelets to the base of the branches.

Spikelets: Laterally compressed, lanceolate-elliptic, florets 3–5, disarticulating above the glumes and between the florets.

> **Glumes:** Glabrous and awnless.
>
> **Lemmas:** Glabrous and awnless.
>
> **Paleas:** About 3/4 the length of the lemmas, membranous, glabrous.
>
> **Caryopses:** Fusiform, brown.

General Comments: Introduced. *Eleusine indica* is a warm-season species that tolerates vehicular and pedestrian traffic. It is difficult to remove by hand weeding.

Mediterranean Lovegrass

Eragrostis barrelieri
J. Daveau

Growth Habit: Annual; tufted.

Culms: Erect or geniculate, branching above the base.

Leaves:

    Sheaths: Margins glabrous but with pilose hairs near the base of the blade.

    Ligules: A short ring of hairs.

    Blades: Flat, glabrous below but with scattered hairs on the upper epidermis.

Inflorescence: A narrow panicle; branches bearing irregularly spaced glandular spots on the pedicels and in the branch axils, not bearing spikelets to the base of the branches.

Spikelets: Lanceolate with 2–12 florets.

    Glumes: Glabrous and awnless.

    Lemmas: Lanceolate-ovate, glabrous, awnless.

    Paleas: About 3/4 the length of the lemmas, membranous.

    Caryopses: Ovate, brown.

General Comments: Introduced. This species is common during the fall in playgrounds; on roadsides; and in other dry, disturbed sites.

Stalkless Lovegrass, Tumble Lovegrass

Eragrostis sessilispica
S. Buckley

Growth Habit: Perennial; tufted.

Culms: Erect, but inflorescence segment curved.

Leaves:

> **Sheaths:** About as long as the internodes.

> **Ligules:** A ring of hairs.

> **Blades:** Involute, glabrous below and with scattered pilose hairs on the upper epidermis.

Inflorescence: Disarticulating at the base of the peduncle and becoming a tumbleweed. Branches with tufts of pilose hairs in the axils, spikelets widely spaced and appressed, spikelets present to the base of the branches.

Spikelets: Linear, florets 8–12.

> **Glumes:** Lanceolate, purple tinged, about the same length, glabrous, awnless.

> **Lemmas:** Lanceolate, purple tinged, glabrous, awnless.

> **Paleas:** About as long as the lemmas, bowed at the base.

> **Caryopses:** Ovoid, brown.

General Comments: Native.

Red Sprangletop

Leptochloa panicea (A. Retzius)
J. Ohwi subsp. *brachiata*
(E. von Steudel) N. Snow

Syn. *L. filiformis* (J. de Lamarck)
A. Palisot de Beauvois

Growth Habit: Annual.

Culms: Erect or geniculate.

Leaves:

> **Sheaths:** Overlapping, glabrous.
>
> **Ligules:** A fringed, glabrous membrane.
>
> **Blades:** Glabrous.

Inflorescence: Lanceolate, branches 1 per node bearing spikelets to the base.

Spikelets: Oblanceolate, florets 2–3, glabrous.

> **Glumes:** About the same length, glabrous.
>
> **Lemmas:** Ovate-lanceolate, as long as the glumes, glabrous, apex bifid.
>
> **Paleas:** Glabrous.
>
> **Caryopses:** Lanceolate-ovate, brown.

General Comments: Native. This species forms dense stands in agricultural fields, particularly in irrigated areas.

Dallisgrass

Paspalum dilatatum
J. Poiret

Growth Habit: Perennial; tufted.

Culms: Geniculate, flattened near the base.

Leaves:

> **Sheaths:** Overlapping, glabrous.
>
> **Ligules:** A fringed, glabrous membrane.
>
> **Blades:** Flat and glabrous.

Inflorescence: A panicle with 1-sided racemose branches, branches glabrous with a tuft of hairs in the axils; branches 1 per node; spikelets arranged in several rows on the lower side of the rachis.

Spikelets: Ovate-lanceolate, florets 2, the lowermost sterile.

> **Glumes:** First glume absent, second glume as long as the spikelet, finely pubescent, margins ciliate pubescent.
>
> **Lemmas:** Lemma of the sterile floret similar to that of the second glume; lemma of the fertile floret ovate-lanceolate, lustrous, indurate.
>
> **Paleas:** Palea of the fertile floret about as long as the lemma.
>
> **Caryopses:** Enclosed within the lemma of the fertile floret.

General Comments: Introduced. Dallisgrass is often introduced into St. Augustine grass lawns with sod. Once established, it is difficult to eradicate. It is occasionally found on mesic roadsides in urban areas. Dallisgrass is more abundant in central and southeastern Texas.

Buffelgrass, African Foxtail

Pennisetum ciliare (C. Linnaeus) J. Link

Syn. *Cenchrus ciliaris* C. Linnaeus

Growth Habit: Perennial; tufted.

Culms: Erect with hard, knotlike bases.

Leaves:

>**Sheaths:** Nearly as long as the internodes, glabrous above but pilose near the base.

>**Ligules:** Membranous with a ring of hairs at the apex.

>**Blades:** Flat with scattered pustulate-base hairs above and glabrous on the lower epidermis.

Inflorescence: A spicate raceme, rachis scabrous or pubescent.

Spikelets: Subtended by a dense cluster of purple bristles and soft hairs.

>**Glumes:** Lanceolate, membranous, glabrous, awnless.

>**Lemmas:** Lanceolate, as long as the spikelet, glabrous, awnless.

>**Paleas:** Membranous, enclosed by the palea.

>**Caryopses:** Ovoid.

General Comments: Introduced. Buffelgrass was introduced from South Africa into the San Antonio area in 1946 (Holt 1985). It has become the most important forage grass in the drier areas of South Texas and northern Mexico, but it is also one of the most aggressive invasive weedy species. It displaces native species and carries wildfires. It recovers rapidly after rain. Buffelgrass is one of the most troublesome weeds in Big Bend National Park (Michael Powell, pers. comm.).

Common Reed

Phragmites australis
(A. Cavanilles)
K. von Trinius *ex*
E. von Steudel

Syn. *P. communis*
K. von Trinius

Growth Habit: Perennial; forming large colonies from rhizomes and stolons.

Culms: Erect, up to 4 m tall, glabrous, hollow at the internodes.

Leaves:

> **Sheaths:** Overlapping, glabrous.
>
> **Ligules:** A ring of hairs.
>
> **Blades:** Flat, up to 2 cm wide, glabrous.

Inflorescence: A large panicle, peduncle with a tuft of hairs at the base.

Spikelets: Laterally compressed, oblanceolate, florets 5–8, rachilla with long, silky hairs.

> **Glumes:** Lanceolate, brown, awnless.
>
> **Lemmas:** Linear-lanceolate, brown, glabrous, awnless.
>
> **Paleas:** Membranous, glabrous.
>
> **Caryopses:** Inconspicuous.

General Comments: Native. Common reed is a cosmopolitan species found on every continent except Antarctica. It is an invasive species that forms extensive colonies along the Rio Grande and canal banks. It is similar in appearance to *Arundo donax*. However, the spikelet rachilla is glabrous in *A. donax* and pubescent in *P. australis.* Common reed is a management problem in some areas of the United States.

Annual Bluegrass

Poa annua C. Linnaeus

Growth Habit: Annual.

Culms: Erect or geniculate, low growing.

Leaves:

> **Sheaths:** About the same length as the internodes, glabrous.
>
> **Ligules:** Membranous, glabrous, broadly triangular at the apex.
>
> **Blades:** Flat, glabrous.

Inflorescence: A panicle with 2 branches per node.

Spikelets: Laterally compressed, linear or lanceolate, florets 4–5, disarticulation above the glumes and between the florets.

> **Glumes:** Lanceolate, glabrous, awnless.
>
> **Lemmas:** Mostly glabrous but pubescent on the lower portion of the keel, awnless.
>
> **Paleas:** Membranous, glabrous or pubescent near the base.
>
> **Caryopses:** Inconspicuous.

General Comments: Introduced. *Poa annua* is a low-growing, cool-season grass that occurs in lawns and moist, shaded sites in urban landscapes. It is a common European weed. It is more common in the moist, temperate areas of Texas than in semiarid areas of the state.

Pathogens and Pests: Two common turf grass diseases are often found on *P. annua*. These include downy mildew (*Sclerophthora macrospora*) and brown patch (*Rhizoctonia solani*). Many varieties of St. Augustine grass, bermudagrass, and zoysia (*Zoysia japonica*) are susceptible to brown patch. See Appendix 5.

Bristly Foxtail

Setaria verticillata
(C. Linnaeus) A. Palisot
de Beauvois var.
respiciens (A. Richard)
A. Braun

Syn. *S. adhaerans*
(P. Forsskål)
E. Chiovenda

Growth Habit: Annual.

Culms: Geniculate,
branched above the base.

Leaves:

> **Sheaths:** Open,
> glabrous.
>
> **Ligules:** A fringed
> membrane with a
> ring of hairs above.
>
> **Blades:** Flat, nearly
> glabrous but with
> scattered pilose
> hairs, margins
> pubescent near the
> base.

Inflorescence: A contracted or racemose panicle, linear, pedicels with 1 retrorsely barbed bristle below each spikelet, spikelets densely overlapping on the axis.

Spikelets: Ovate-elliptic.

Glumes: Ovate or ovate-elliptic, apex obtuse.

Lemmas: Lemma of the fertile floret, indurate, finely rugose, glabrous, awnless.

Paleas: About as long as the lemma of the fertile floret.

Caryopses: Enclosed within the fertile floret.

General Comments: Introduced. This warm-season species is common in flower beds, gardens, and recently cultivated sandy loam fields.

Johnsongrass

Sorghum halepense
(C. Linnaeus) C. Persoon

Growth Habit: Perennial; rhizomatous.

Culms: Erect or geniculate, glabrous but finely pubescent at the nodes.

Leaves:

> **Sheaths:** Glabrous.

> **Ligules:** A fringed, slightly pubescent membrane.

> **Blades:** Flat, glabrous.

Inflorescence: An open panicle, rachis pubescent.

Spikelets: Paired, 1 sessile and perfect and 1 pedicelled and staminate or neuter, ovate-lanceolate, disarticulation below the spikelet pairs, pedicels flattened with dense, pilose hairs.

> **Glumes:** Lanceolate, as long as the spikelet, yellow-purple, pubescent.

> **Lemmas:** Membranous, margins pilose.

> **Paleas:** Enclosed by the lemma.

> **Caryopses:** Enclosed within the fertile floret.

General Comments: Introduced. *Sorghum halepense*, originally from Asia, has colonized most temperate and subtropical latitudes (Dahlberg 2000). It is most commonly referred to as johnsongrass. However, there is a species with thicker culms, larger spikelets, and larger panicles identified as *S. miliaceum* that is often misidentified as johnsongrass. *Sorghum halepense* also commonly hybridizes with cultivated sorghum. Therefore, there is a possibility of a large amount of morpho-logical diversity, especially in sorghum-growing regions (Dahlberg 2000). Johnson-grass reproduces rapidly from rhizomes. Sperry et al. (1968) and Hart et al. (2003) indicate that it produces glycosides that are toxic to livestock under certain conditions. Resistance to triazine herbicides is quite common in johnsongrass throughout the United States.

Pathogens and Pests: Refer to Appendix 6 for an extensive list of pathogens and insect pests of *S. halepense*.

Whorled Dropseed

Sporobolus pyramidatus
(J. de Lamarck)
A. S. Hitchcock

Growth Habit: Perennial; tufted.

Culms: Erect or geniculate, low growing.

Leaves:

> **Sheaths:** About 3/4 the length of the internodes.

> **Ligules:** A ring of hairs.

> **Blades:** Flat or rolled, glabrous, margins with widely scattered serrations.

Inflorescence: A panicle with whorled lower branches, branches arising at about 45-degree angle to the axis, branches 6–8 per node.

Spikelets: Lanceolate or oblanceolate, florets 1, disarticulating above the glumes.

> **Glumes:** Lanceolate, membranous.

> **Lemmas:** Lanceolate, as long as the spikelet, translucent.

> **Paleas:** Membranous, glabrous.

> **Caryopses:** Ovoid.

General Comments: Native. Whorled dropseed and *Heliotropium curassavicum* (seaside heliotrope) are often abundant on the margins of salty areas in agricultural fields and are indicator species of salinization.

Browntop Liverseed Grass

Urochloa fasciculata
(O. Swartz)
R. D. Webster

Syn. *Brachiaria fasciculata* (O. Swartz) L. Parodi; *Panicum fasciculatum* O. Swartz

Growth Habit: Annual.

Culms: Erect or geniculate, branched above the base, glabrous.

Leaves:

> **Sheaths:** About as long as the internodes, glabrous.
>
> **Ligules:** Membranous with a ring of hairs at the apex.
>
> **Blades:** Flat, glabrous, margins minutely scabrous.

Inflorescence: A panicle, branches 1 per node, peduncle scabrous with pilose hairs in the axis.

Spikelets: Nearly round in cross section, pedicels scabrous with scattered pilose hairs.

> **Glumes:** Brownish-green, glabrous with reticulate venation, awnless.
>
> **Lemmas:** Lemmas of the sterile floret similar to that of the second glume, reticulate-veined; lemma of the fertile floret yellow, indurate, rugose, lanceolate-ovate.
>
> **Paleas:** Palea of the fertile floret partially enclosed by the lemma.
>
> **Caryopses:** Ovoid, yellow.

General Comments: Native. This species is abundant in agricultural fields during the warm season.

Guineagrass

Urochloa maxima (N. von Jacquin)
R. D. Webster

Syn. *Panicum maximum*
N. von Jacquin

Growth Habit: Perennial; tufted, robust.

Culms: Erect, unbranched, up to 2 m tall, glabrous.

Leaves:

> **Sheaths:** About 2/3 the length of the internodes, glabrous; margins glabrous below, appressed pubescent above.

> **Ligules:** A fringed membrane with hairs at the apex.

> **Blades:** Flat or occasionally folded, mostly glabrous but with a few appressed hairs above; margins serrate.

Inflorescence: An open panicle, with numerous whorled branches at the base; spikelets absent at the base of the branches.

Spikelets: Elliptic-lanceolate, pedicels antrorsely scabrous.

> **Glumes:** Purple tinged, apex blunt or obtuse.

> **Lemmas:** Lemma of the sterile floret similar to that of the second glume; lemmas of the fertile floret lanceolate-elliptic, whitish, transversely rugose.

> **Paleas:** Partially enclosed by the lemma of the fertile floret.

> **Caryopses:** Ovate, yellow.

General Comments: Introduced. Guineagrass is a major weed control problem in more mesic areas of South Texas. In the Rio Grande riparian corridor this invasive species excludes nearly all of the native herbaceous species (Lonard and Judd 2002). It presents a fire hazard during drought conditions. It recovers quickly after fire and drought.

Panic Liverseed Grass

Urochloa panicoides
A. Palisot de Beauvois

Growth Habit: Annual; long-lived, rooting at the lower nodes.

Culms: Prostrate or ascending from a stoloniferous base, freely branching, glabrous or with a few long, straight hairs.

Leaves:

>**Sheaths:** About 1/2 the length of the internodes with scattered straight hairs; margins crisped with long, straight hairs and antrorsely angled serrations.

>**Ligules:** A ring of bristly hairs.

>**Blades:** Flat with scattered, straight, papillose-based hairs and antrorsely scabrous prickles.

Inflorescence: A panicle with 1-sided racemose branches; branches 1 per node, peduncles with a dense tuft of hairs at the base.

Spikelets: Nearly round in cross section, ovate, florets 2, the uppermost fertile, pedicels with 1–3 long hairs.

>**Glumes:** Glabrous, awnless; first glume and lemma of the sterile floret oriented away from the rachis axis.

>**Lemmas:** Lemmas of the sterile floret tapered to a sharp point; lemmas of the fertile floret yellow, awned with rugose papillae.

>**Paleas:** Palea of the sterile floret about as long as the lemma, membranous; palea of the fertile floret as long as the lemma, bony.

>**Caryopses:** Ovate, yellow or brown, about as long as the palea.

General Comments: Introduced. This species is a federally listed noxious weed that is spreading rapidly in the warmer areas of Texas. It occurs on roadsides, in lawns, on playgrounds, and in a wide variety of other disturbed sites. It tolerates heavy foot traffic and moderate vehicular traffic. It causes nitrate poisoning in livestock.

Creeping Liverseed Grass

Urochloa reptans (C. Linnaeus)
O. Stapf

Syn. *Brachiaria reptans* (C. Linnaeus)
C. A. Gardner & C. Hubbard; *Panicum reptans* C. Linnaeus

Growth Habit: Annual; mat forming, stoloniferous, rooting at the nodes.

Culms: Prostrate, upper culms becoming erect, branching freely, glabrous but with a ring of hairs at the nodes.

Leaves:

> **Sheaths:** As long as the internodes, glabrous; margins densely hirsute.
>
> **Ligules:** A ring of hairs.
>
> **Blades:** Flat, glabrous above but with a few scattered hairs on the lower epidermis, crisped on the margins and antrorsely scabrous.

Inflorescence: A panicle with 1-sided racemose branches, branches 1–2 per node, rachis antrorsely serrate.

Spikelets: Paired, 1 on a longer pedicel, ovate; pedicels with scattered, long hairs.

> **Glumes:** Ovate-elliptic, glabrous, awnless.
>
> **Lemmas:** Lemma of the sterile floret ovate, glabrous, awnless; lemma of the fertile floret ovate, glabrous, white, awn tipped.
>
> **Paleas:** Palea of the sterile floret membranous; palea of the fertile floret indurate, rugose.
>
> **Caryopses:** Enclosed by the fertile floret.

General Comments: Introduced. Creeping liverseed grass is a warm-season weed that is common in flower beds. It spreads rapidly over bare soil.

Texas Liverseed Grass

Urochloa texana
(S. Buckley)
R. D. Webster

Syn. *Brachiaria texana*
(S. Buckley) S. T. Blake;
Panicum texanum
S. Buckley

Growth Habit: Annual; tufted, often rooting at the lower nodes.

Culms: Erect or geniculate, pubescent.

Leaves:

    Sheaths: About as long as the internodes, pubescent.

    Ligules: A fringed membrane with a ring of hairs at the apex.

    Blades: Flat, 1–2 cm wide, pubescent; margins crisped-serrate, pubescent near the base.

Inflorescence: A contracted panicle, branches 1–2 per node; peduncles hirsute.

Spikelets: Ovate-lanceolate, nearly round in cross section, pedicels hirsute.

    Glumes: Glabrous, awnless.

    Lemmas: Awnless; lemma of the sterile floret similar to that of the second glume; lemma of the fertile floret rugose, white with a mucronate apex.

    Paleas: Palea of the fertile floret rugose.

    Caryopses: About 2 mm long.

General Comments: Native. Texas liverseed grass is abundant in cultivated fields during the warm season.

PONTEDERIACEAE

Waterhyacinth

Eichhornia crassipes
(K. von Martius)
C. Solms-Lauchach

Growth Habit: Perennial; free floating.

Stems: Aerial stems absent, stolons present.

Leaves: Simple, in a rosette; blades suborbicular or broadly elliptic, subtended by a sheath, bright green, glossy, petioles inflated.

Inflorescence: A spike bearing several to numerous showy flowers.

> **Calyx and Corolla:** Sepals and petals similar in size and appearance (tepals), slightly zygomorphic, light blue to purplish-blue.
>
> **Stamens:** 6.
>
> **Pistil:** Ovary superior, style 1.

Fruit: A capsule; seeds minute, abundant.

General Comments: Introduced. Waterhyacinth is one of the world's worst aquatic weeds. It is a native of South America that is now found in many tropical and subtropical areas of the world. It proliferates rapidly and forms large, free-floating rafts. It can double the area it covers in two weeks. Through the process of transpiration, the rate of water lost to the atmosphere in areas inundated with waterhyacinth may be 4 to 5 times as great as in areas with open water (Mitchell 1976). Waterhyacinth is a serious weed problem in several South Texas waterways, including canals and the Rio Grande in the LRGV and Lake Corpus Christi. Recommended control methods include mechanically removing, shredding, or using the herbicide 2,4-D (Anonymous 2001).

TYPHACEAE

Southern Cattail

Typha domingensis
C. Persoon

Growth Habit: Perennial; from an extensive rhizome system; forming large colonies; plants monoecious.

Stems: Erect, glabrous.

Leaves: Simple, alternate; blades strap shaped, glabrous, sheathing the stem, flat to slightly rounded below, venation parallel and lacking a distinct midvein.

Inflorescence: An interrupted spike, the staminate flowers above the pistillate; a bare zone separates the staminate and pistillate flowers.

> **Staminate Flowers:** Perianth reduced to minute bracts; inflorescence covered with a dense mass of yellow pollen; flowers falling after pollen production, leaving a persistent, naked rachis.

> **Pistillate Flowers:** Perianth minute, brownish, persistent on the rachis.

Fruit: Minute, plumed nutlets.

General Comments: Native. Southern cattail forms extensive colonies from rhizomes that clog irrigation canals and shallow freshwater and brackish wetlands.

Appendices

The information contained in these tables is partial and represents only a selected portion of the known pathogens and pests that have been encountered on a particular plant species. For more information concerning any crop, weed, pathogen, or pest, please consult your county extension agent.

APPENDIX 1

Toxic Plants and Federally Listed Noxious Weeds of South Texas

Toxic Plants

Amaranthus blitoides

A. palmeri

A. rudis

Chamaesyce prostrata

Conyza canadensis

Delphinium carolinianum var. *virescens*

Descurainia pinnata

Gutierrezia sarothrae

Helenium microcephalum

Hymenoxys odorata

Kallstroemia californica

Lobelia berlandieri

Melilotus albus

Ricinus communis

Rumex chrysocarpus

Salsola tragus

Senecio ampullaceus

Solanum americanum

S. elaeagnifolium

S. rostratum

Sorghum halepense

Trianthema portulacastrum

Tribulus terrestris

Urochloa panicoides

Verbesina encelioides

Xanthium strumarium

Federally Listed Noxious Weeds

Salvinia molesta

Tridax procumbens

Urochloa panicoides

APPENDIX 2

Selected Pathogens and Insect Pests of Sunflower (*Helianthus annuus*) in Texas

Pathogen or Disease or Pest Type	Common Name	Pathogen or Pest Name	Notes
Virus	Sunflower mosaic	SuMV	SuMV is a potyvirus that causes mosaic symptoms in many plants of the Asteraceae; SuMV has been collected from wild sunflowers in Cameron and Hidalgo counties in South Texas.
Fungus	Charcoal rot	*Macrophomina phaseolina*	Root and stem rot.
	Downy mildew	*Plasmopara halstedii*	*P. halstedii* causes downy mildews of members of the Asteraceae, including *Ambrosia* spp. and *Bidens* spp.
	Head rot	*Rhizopus* spp.	Head rot species include *Rhizopus arrhizus*, *R. oryzae*, and *R. stolonifer*.
	Leaf spots	*Cercospora pachypus*	
	Powdery mildew	*Erysiphe cichoracearum*	*E. cichoracearum* has also been reported on *Cucumis melo* in the LRGV and on *Lycopersicon esculentum*, *Senecio* spp., *Citrullus* spp., and *Cucurbita* spp. in Mexico.
	Root rots	*Phymatotrichopsis omnivora*	*P. omnivora* causes "cotton root rot" (a.k.a. *Phymatotrichum omnivorum*).
		Rhizoctonia crocorum	*R. crocorum* causes "violet root rot" in cultivated sunflowers.
		R. solani	*R. solani* has been reported to also cause disease on *Daucus carota*, *Gossypium hirsutum*, and *Hibiscus cannabinus* in the LRGV. *R. solani* has also been reported on *Mangiferae indica*, *Medicago sativa*, and *Prunus amygdalus* in Mexico.
	Rust	*Puccinia helianthi*	Stages 0, I, II, III, and IV of *P. helianthi* occur on sunflower.*
	Sclerotinia wilt	*Sclerotinia sclerotiorum*	
	Southern blight	*Sclerotium rolfsii*	*S. rolfsii* is also reported on *Sorghum bicolor*, where it can cause a leaf/sheath blight.
	Stem canker	*Diaporthe* spp.†	
	Wilt	*Fusarium* spp., *Verticillium* sp.	
Nematode	Citrus nematode	*Tylenchulus semipenetrans*	*T. semipenetrans* is also commonly found on grapefruit in the LRGV.
	Reniform nematode	*Rotylenchulus reniformis*	*R. reniformis* is found on cotton (*Gossypium hirsutum* and *G. barbadense* in the LRGV.
	Root-knot nematode	*Meloidogyne incognita*	Common pathogenic nematode on many cultivated plants in Texas.

Pathogen or Disease or Pest Type	Common Name	Pathogen or Pest Name	Notes
Arthropod	Beet armyworm	*Spodoptera exigua*	*S. exigua* is also reported on cotton (*Gossypium* spp.), cabbage (*Brassica oleracea*), bell pepper (*Capsicum annuum*), and pigweed (*Amaranthus* spp.) in the LRGV.
	Corn earworm	*Helicoverpa zeae*	Sunflower can be injured by the larvae, though corn and lettuce (*Lactuca* spp.) are the best hosts in South Texas.
	Leaf miner	*Liriomyza* spp.	
	Lubber grasshopper	*Brachystola magna*	
	Pearl crescent	*Phycioides tharos*	
	Sunflower moth	*Homoesoma electellum*	*H. electellum* is associated with the *Rhizopus* head rot disease of cultivated sunflowers.
	Threecornered leafhopper	*Spissistilus festinus*	
	Tumbling flower beetle	*Mordella* spp.	

*Rust pathogen stages: 0 = spermatogonium (pycnium)/spermatia (pycniospores); I = aecium/aeciospores; II = uredinium/urediniospores; III = telium/teliospores; IV = probasidium/basidiospores.

†Stem canker and leaf spot caused by *D. helianthi* in Texas.

References: Alvarez (1976); Cook and Riggs (1993); Crawford (1970); Cummins (1964); Davis, Heald, and Timmer (1982); Drees and Jackman (1998); Farr et al. (1989); Godfrey (1952); Graham and Robertson (1970); Greenberg et al. (2001); Gulya et al. (2002); Horne (1988); Milhail (1992); Odvody and Madden (1984); Orellana (1971); Robinson et al. (1987); Rogers, Thompson, and Zimmer (1978); Rush, Gerik, and Kenerley (1985); Spring et al. (2003); Westphal and Smart (2003); Yang and Owen (1982); Yang et al. (1979); Yang et al. (1984).

APPENDIX 3

Selected Pathogens and Insect Pests of Giant Reed (*Arundo donax*) in Texas

Pathogen or Disease or Pest Type	Common Name	Pathogen or Pest Name	Notes
Virus	Mosaic	Maize dwarf mosaic virus	MDMV is Aphididae transmitted.
		Sugarcane mosaic virus	SCMV is a potyvirus that is also seen on maize, sorghum, and sugarcane. SCMV will not infect johnsongrass.
Fungus	Panicle blight	*Gibberella fujikuroi* mating population D	The asexual form of *G. fujikuroi* mating population D is *Fusarium proliferatum*.
	Frogeye	*Septoria donacis*	
	Leaf spots	*Papularia sphaerospermum*	
		Scolecotrichum maculicola	
Arthropod	Greenbug	*Schizaphis graminum*	*S. graminum* uses *A. donax* as a winter feeding source.
	Sugarcane borer	*Diatraea saccharalis*	

References: Cowan (1962); Drees and Jackman (1998); Farr et al. (1989); Kerenyi et al. (2000); Koike and Gillaspie (1989); Sprague (1950); Tucker (1940).

APPENDIX 4

Selected Pathogens and Insect Pests of Bermudagrass (*Cynodon dactylon*) in Texas

Pathogen or Disease or Pest Type	Common Name	Pathogen or Pest Name	Notes
Fungus	Anthracnose	*Colletotrichum graminicola*	*C. graminicola* has also been reported on *Bromus catharticus, Sorghum halepense, S. bicolor,* and *Zea mays.*
	Bipolaris leaf spot	*Bipolaris* spp.	*Bipolaris* spp. may also cause crown and root rots on *Cynodon dactylon.* Common species of *Bipolaris* include *B. cynodontis* and *B. stenospila.* Similar disease also caused by other fungi, including *Curvularia, Dreschlera,* and *Exserohilum.*
	Brown patch	*Rhizoctonia* spp.	*Rhizoctonia* species on *Cynodon dactylon* include *R. oryzae, R. solani,* and *R. zeae. R. solani* has been reported to also cause disease on *Daucus carota, Gossypium hirsutum,* and *Hibiscus cannabinus* in the LRGV. *R. solani* has also been reported on *Mangiferae indica, Medicago sativa,* and *Prunus amygdalus* in Mexico.
	Crown rust	*Puccinia coronata*	*P. coronata* has been reported on *Bromus* spp. in Mexico. Stages 0 and I occur on *Rhamnus* spp., but other members of Rhamnaceae have not been identified as alternative hosts. Therefore, it is unlikely that this rust is prevalent or important on bermudagrass in South Texas.
	Gray leaf spot (Blast)	*Magnaporthe grisea*	*M. grisea* causes blast in rice and a number of other cereal and grass species.
	Leaf blotch	*Bipolaris cynodontis*	Also known as Helminthosporium leaf blotch.
	Leaf rust	*Puccinia* spp.	Factors promulgating leaf rust disease: 70°–75°F temperatures (therefore this disease is only rarely seen in South Texas and northern Mexico in late fall and early spring), nitrogen deficiency, drought stress, short clipping. Species in South Texas include *P. brachypodii* and *P. cynodontis.*
	Leaf spots	*Aschochyta graminae*	
		Cercospora seminalis	*C. seminalis* causes a disease that has also been referred to as "false smut" (may cause a glume blotch that resembles a smutted spikelet).
		Dreschlera gigantea	*D. gigantea* has been classically referred to as *Helminthosporium giganteum* and causes "zonate eyespot."
		Leptostromella cynodontis	*L. cynodontis* causes a decline in hybrid (lawn) bermudagrass cultivars.
		Septoria cynodontis	

Pathogen or Disease or Pest Type	Common Name	Pathogen or Pest Name	Notes
Fungus	Necrotic ring spot	*Leptosphaeria korrae*	A complex of fungi may cause necrotic ring spot. Other fungal pathogens include *Ophiosphaerella herpotricha* and *Gaeumannomyces graminis*.
	Powdery mildew	*Blumeria* (*Erysiphe*)	*B. graminis* has also been reported on *Bromus catharticus* and *Digitaria graminis ciliaris*.
	Pythium blight (Cottony blight)	*Pythium* spp.	Factors promulgating disease: heat stress (>85°F), poor soil drainage (excessive moisture). An older disease name, "greasy spot," has been used for this disease, but its use should be avoided in South Texas and northern Mexico because it is a commonly used disease name for a defoliating citrus disease (caused by *Mycosphaerella citri*). *Pythium* spp. responsible for this disease include *P. aphanidermatum, P. graminicola, and P. ultimum*. *P. graminicola* has been reported from sorghum and corn. *P. ultimum* causes seedling damping-off on a wide range of vegetables and row crops.
	Seedling blight	*Sclerotium rolfsii*	This fungus causes the common plant disease known as "southern blight." This fungus is also associated with a basal sheath and stalk rot of *Cynodon dactylon*.
	Smut	*Sporisorium cenchri*	*S. cenchri* causes an "inflorescence smut."
		S. sorghi	*S. sorghi* is also a pathogen of sorghum. *C. dactylon* acts as an alternative host for *S. sorghi*.
		Ustilago cynodontis	*U. cynodontis* is very predominant on bermudagrass in South Texas.
	Spring dead spot	*Gaeumannomyces graminis*	This disease can be differentiated from "take-all" and bermudagrass "decline," although these are caused by the same pathogen.
	Stem rust	*Puccinia graminis*	Alternative hosts include *Berberis* spp. and *Mahonia* spp.
Nematode	Cyst nematode	*Heterodera* spp.	
	Dagger nematode	*Xiphenema* spp.	Example: *X. americanum;* also a pathogen of corn and various grasses.
	Lance nematode	*Hoplolaimus* spp.	Example: *H. tylenchiformis*.
	Needle nematode	*Longidorus* spp.	
	Pin nematode	*Paratalynchus* spp.	
	Root-knot nematode	*Meloidogyne* spp.	Example: *M. incognita*.
	Spiral nematode	*Helicotylenchus* spp.	
	Sting nematode	*Belonolaimus* spp.	Example: *B. longicaudatus,* also a pathogen of corn and various grasses.
	Stunt nematode	*Tylenchorhynchus* spp.	
	Stubby root nematode	*Paratrichadorus* spp., *Trichodorus* spp.	Example: *P. christei.*

Pathogen or Disease or Pest Type	Common Name	Pathogen or Pest Name	Notes
Arthropod	Bermudagrass mite	*Eriophyes cynodoniensis*	
	Chinchbug	*Blissus* spp.	
	Grubworm	Many species	
	Masked chafer	*Cyclocephala* spp.	
	Sod webworm	Many species	
	Threecornered alfalfa hopper	*Spissistilus festinus*	

References: Alvarez (1976); Browning et al. (1999); Cowan (1962); Cummins (1964); Cummins and Greene (1966); Drees and Jackman (1998); Farr et al. (1989); Fischer (1953); Horne (1988); Horvath and Vargas (2004); Marley (1995); Shurtleff (1980); Smiley, Dernoeden, and Clarke (2005); Sprague (1950); Wilson (1999).

APPENDIX 5

Selected Pathogens and Insect Pests of Annual Bluegrass (*Poa annua*) in Texas

Pathogen or Disease or Pest Type	Common Name	Pathogen or Pest Name	Notes
Fungus	Bipolaris leaf spot	*Bipolaris sorokiana*	The sexual stage of *B. sorokiana* is *Cochliobolus sativus*.
	Brown patch	*Rhizoctonia* spp.	*Rhizoctonia* species on *P. annua* include *R. oryzae, R. solani,* and *R. zeae. R. solani* has been reported to also cause disease on *Daucus carota, Gossypium hirsutum,* and *Hibiscus cannabinus* in the LRGV. *R. solani* has also been reported on *Mangiferae indica, Medicago sativa,* and *Prunus amygdalus* in Mexico.
	Crown rust	*Puccinia coronata*	*P. coronata* has been reported on *Bromus* spp. in Mexico. Stages 0 and I occur on *Rhamnus* spp., but other members of Rhamnaceae have not been identified as alternative hosts. Therefore, it is unlikely that this rust is prevalent or important on bluegrass in South Texas.
	Leaf rust	*Puccinia* spp.	Factors promulgating leaf rust disease: 70°–75°F temperatures (therefore this disease is only rarely seen in South Texas and northern Mexico in late fall and early spring), nitrogen deficiency, drought stress, short clipping. Species in South Texas include *P. brachypodii* and *P. cynodontis*.
	Leptosphaerulina blight	*Leptosphaerulina trifolii*	
	Necrotic ring spot	*Leptosphaeria korrae*	
	Powdery mildew	*Blumeria (Erysiphe) graminis*	*B. graminis* has been reported on *Bromus catharticus* and *Digitaria ciliaris*.
	Pythium blight (Cottony blight)	*Pythium* spp.	Factors promulgating pythium blight: heat stress (>85°F), poor soil drainage (excessive moisture). An older disease name, "greasy spot," has been used for this disease, but its use should be avoided in South Texas because it is similar to a commonly used disease name for a defoliating citrus disease (caused by *Mycosphaerella citri,* e.g., "greasy spot"). *Pythium* spp. responsible for this disease include *P. aphanidermatum, P. graminicola,* and *P. ultimum. P. graminicola* has been reported from sorghum and corn. *P. ultimum* causes seedling damping-off of a wide range of vegetables and row crops.
	Summer patch	*Magnaporthe poae*	A related species, *M. grisea,* causes blast in rice and a number of other cereals and grasses.
	Take-all	*Gaeumannomyces graminis*	Take-all disease is characterized by "whiteheads" composed of sterile spikelets in the flowering grass as well as extensive rotting and rhizomorphs found on plant roots.

Pathogen or Disease or Pest Type	Common Name	Pathogen or Pest Name	Notes
Nematode	Cyst nematode	*Heterodera* spp.	
	Dagger nematode	*Xiphenema* spp.	
	Lance nematode	*Hoplolaimus tylenchiformis*	
	Needle nematode	*Longidorus* spp.	
	Pin nematode	*Paratalynchus* spp.	
	Root-knot nematode	*Meloidogyne* spp.	
	Spiral nematode	*Helicotylenchus* spp.	
	Sting nematode	*Belonolaimus* spp.	
	Stubby root nematode	*Paratrichadorus, Trichodorus* spp.	
	Stunt nematode	*Tylenchorhynchus claytoni*	
Arthropod	Chinchbug	*Blissus* spp.	
	Greenbug	*Schizaphis graminum*	
	Masked chafer (grub)	*Cyclocephala* spp.	
	Sod webworm	*Herpetogramma phaeopteralis*	

References: Alvarez (1976); Cowan (1962); Cummins (1964); Cummins and Greene (1966); Drees and Jackman (1998); Gibb and Buhler (1995); Horne (1988); Orellana (1971); Schrock (2004); Smiley, Dernoeden, and Clarke (2005); Sprague (1950).

APPENDIX 6

Selected Pathogens and Insect Pests of Johnsongrass (*Sorghum halepense*) in Texas

Pathogen or Disease or Pest Type	Common Name	Pathogen or Pest Name	Notes
Virus	Johnsongrass mosaic virus	JgMV (Potyvirus)	Two other potyviruses infect sorghum but *not* johnsongrass: sugarcane mosaic virus (SCMV) and sorghum mosaic virus (SrMV).
	Maize chlorotic dwarf virus	MCDV	Occurs on millets, crabgrass, maize, sorghum, sudangrass, and wheat. Perennial overwintering host; vectors: leafhoppers *Graminella nigrifons* and *G. sonora*.
	Maize dwarf mosaic virus	MDMV (Potyvirus)	Occurs on maize, sorghum, and other perennial grasses. *S. halepense* is a perennial overwintering host; vectors: greenbug, green leaf aphid, green peach aphid; at least 12 species of aphids in total.
Fungus	Anthracnose	*Colletotrichum graminicola*	This pathogen is quite hypervariable and may be virulent on a number of cereal and grass species.
	Cottony blight	*Pythium ultimum*	
	Covered kernel smut	*Sporisorium sorghi*	
	Downy mildew	*Peronosclerospora sorghi*	
	Ergot	*Claviceps africana*	This fungus was first observed on sorghum and johnsongrass in the United States in Cameron and Hidalgo counties of South Texas in 1997.
	Leaf spots	*Cercospora sorghi*	*C. sorghi* causes "gray leaf spot" and is a prevalent leaf spot on cultivated sorghum.
		Exserohilum turcicum	*E. turcicum* also causes seedling blights and root rots and is a pathogen of sorghum, where it causes "leaf blight."
		Phoma sorghina	
		Phyllosticta sorghina	
		Septoria pertusa	*S. pertusa* was observed extensively in the Rio Grande valley of New Mexico by F. Weiss (ca. 1910), and it causes leaf spots on sorghum.
	Loose kernel smut	*Sporisorium cruetnum*	The older name for this fungus is *Sphacelotheca cruenta;* this pathogen also infects sorghum.
	Red rot	*Glomerella (Physalospora) tucumanensis*	This fungus is also responsible for a leaf spot and anthracnose disease (anamorph: *Colletotrichum falcatum*).
	Rust	*Puccinia purpurea*	Sorghum and johnsongrass may serve as the uredinial and telial hosts for this rust.
	Target spot	*Bipolaris sorghicola*	
	Zonate leaf spot	*Gloeocercospora sorghi*	

Pathogen or Disease or Pest Type	Common Name	Pathogen or Pest Name	Notes
Nematode	Dagger nematode	*Xiphenema* spp.	
	Lance nematode	*Hoplolaimus* spp.	
	Pin nematode	*Paratalynchus* spp.	
	Root-knot nematode	*Meloidogyne* spp.	
	Spiral nematode	*Helicotylenchus* spp.	
	Sting nematode	*Belonolaimus* spp.	
	Stunt nematode	*Tylenchorhynchus* spp.	
Arthropod	Rice stink bug	*Oebalus pugnax*	
	Sorghum midge	*Contarinia sorghicola*	Early insect stages develop on *S. halepense* before moving to *S. bicolor*.
	Sorghum webworm	*Nola sorghiella*	
	Threecornered alfalfa hopper	*Spissistilus festinus*	
	Yellow sugarcane aphid	*Sipha flava*	Cultivated sorghum and johnsongrass are susceptible to a toxin that is released from the aphid when feeding, resulting in pigment accumulation in colored genotypes.

References: Alderman, Halse, and White (2004); Cowan (1962); Cummins (1964); Dean (1966); Drees and Jackman (1998); Farr et al. (1989); Fischer (1953); Frederiksen and Odvody (2000); Futrell and Frederiksen (1970); Horne (1988); Horvath and Vargas (2004); Isakeit, Odvody, and Shelby (1998); Rosewich et al. (1998); Sprague (1950).

APPENDIX 7

Selected Chemical and Cultural Controls of Some Monocot (M) and Dicot (D) Weeds of South Texas and Northern Mexico

Weed Species	Herbicidal Control	Notes/Cultural Control
Alternanthera caracasana (D)	2,4-D*, Confront®, dicamba, mecoprop	
Ambrosia spp. (D)	2,4-D, dicamba, mecoprop	
Arundo donax (M)	Glyphosate	
Convolvulus spp. (D)	2,4-D, dicamba, mecoprop, quinclorac	Manually remove vines.
Digitaria spp. (M)	DCPA†, fluazifop, glyphosate	Maintain a thick lawn.
Echinochloa spp. (M)	DCPA, fluazifop	
Hydrilla verticilliata (M)	2,4-D	
Lamium amplexicaule (D)	2,4-D, mecoprop, chlorsulfuron	Manually remove plants.
Malva spp. (D)		Maintain a thick lawn.
Myriophyllum sibiricum (D)	2,4-D	
Oxalis stricta (D)	2,4-D, dicamba, mecoprop	Keep soil pH range between 6.5 and 7.7; maintain active lawn growth through regular fertilization.
Poa annua (M)	Bensulide, DCPA	Keep lawn height at 2–4 inches.
Portulaca oleracea (D)	2,4-D, mecoprop, trifluralin, pendimethalin, glyphosate	
Stellaria media (D)	2,4-D, mecoprop	
Taraxacum officinale (D)	2,4-D, atrazine,†† mecoprop	
Trifolium spp. (D)	2,4-D, dicamba, mecoprop	

*(2,4-dichlorophenoxy) acetic acid.

†Chlorthal (tetrachloroterephthalic acid).

††Use only on lawns of St. Augustine grass, as it is deleterious for bermudagrass lawns.

References: Anonymous (1971); Dudley (2000); Smith (1993).

GLOSSARY

achene A small, dry, indehiscent, 1-seeded fruit with a thin ovary wall free from the seed.

actinomorphic Pertaining to a flower approaching radial symmetry.

acuminate Long pointed; tapering to an elongated point.

aeciospore A dikaryotic spore produced by the aecium of a rust fungus.

aecium (aecia) The fruiting structure of a rust fungus most commonly found on the lower epidermis of a leaf. This structure results from the "fertilization" of spermatogonial-receptive hyphae by spermatia on the upper surface of the plant leaf.

alternate Pertaining to a leaf borne singly at each node.

annual A plant that germinates, flowers, produces seeds, and dies in the same year.

anther The expanded, pollen-bearing portion of the stamen.

antrorse Directed upward.

apex The tip; the point farthest from the point of attachment.

appressed Lying flat against a surface.

auricle A small, ear-shaped structure.

auriculate Pertaining to an ear-shaped appendage.

awn Stiff, needlelike extension.

axil The upper angle between the stem and the leaf.

berry A fleshy fruit with more than one seed; the seeds are embedded in pulpy tissue; e.g., a tomato.

biennial Completing the life cycle in two growing seasons.

bifid Forked.

bifurcate Forked or with 2 prongs.

bilabiate Divided into 2 parts or 2 lips.

bipinnately compound Twice-pinnately compound.

blade The expanded, laminar, usually flat portion of a leaf.

bract A reduced or modified leaf often associated with a flower or an inflorescence.

bulb An underground stem with thickened, fleshy scales; e.g., an onion.

calyx Collectively, the sepals of a flower; the outermost series of floral parts of the perianth.

capitate Head shaped.

capsule A simple, dry fruit with several partitions and dehiscing at maturity, several to many seeded.

carpel A simple pistil or one segment of a compound pistil.

caruncle A protuberance near the seed scar; as seen in members of the Euphorbiaceae.

caryopsis A grain or fruit of grasses.

cauline Pertaining to the stem or culm.

chaff A thin, dry scale that arises from the receptacle in some members of the Asteraceae.

chlorosis Yellowing of plant foliage.

circumscissile Pertaining to dehiscence so that the top separates like a lid.

clavate Club shaped.

clawed Pertaining to a narrow, tapered base of some petals and sepals.

cm Centimeter; 0.01 meter.

compound Composed of two or more similar elements; more than one blade per petiole.

connate Having similar organs united as one.

cordate Heart shaped with the notch at the base and ovate in general outline.

corolla Petals of a flower.

corymb A raceme in which the axis is short and the lower pedicels are relatively long; the inflorescence is essentially flat topped.

corymbose Arranged in corymbs.

crenate Pertaining to a leaf margin with blunt or rounded teeth.

crisped Curled.

culm A grass stem.

cyathium An inflorescence consisting of a

cuplike involucre containing a single pistil and male flower with a single stamen; occurs in members of *Chamaesyce* and *Euphorbia*.

cyme An inflorescence in which the central flower of the group is the most mature.

cymose Cymelike.

deciduous Falling off; not persistent.

decumbent Lying on the ground but with the tip ascending.

dehiscent Splitting open at maturity.

diadelphous Pertaining to stamens united into two unequal sets by their filaments, usually in a 9 + 1 arrangement.

dichotomous Forking in pairs.

digitate Arising from one point; fingerlike.

dioecious Pertaining to staminate and pistillate flowers on separate plants.

discoid head A head with ray florets lacking.

disk florets Tubular florets of the Asteraceae.

elliptic Longer than wide and rounded at both ends; pertaining to an ellipse.

enation An outgrowth of the surface of a stem or leaf.

entire Pertaining to an even or smooth margin; lacking teeth.

ephemeral Lasting for a day or less.

epipetalous Pertaining to stamens attached to the corolla.

even-pinnately compound Having terminally paired leaflets.

exserted Protruding; usually pertains to structures extended beyond the length of the corolla.

fascicle A cluster.

fibrous With roots numerous and similar in diameter.

filament The threadlike structure of a stamen that supports the anthers.

floret A unit of a grass spikelet consisting of the lemma, palea, stamens, and pistil or the individual flower in the Asteraceae.

follicle A dry fruit opening on one side of an elongated fruit.

fruit A ripened ovary and associated structures that ripen with the ovary.

fruiting structure A common name for a spore-producing structure of a fungus.

geniculate Abruptly bent.

glabrous Lacking hairs.

glandular Pertaining to a secreting organ, often at the tip of a hair.

glume A pair of sterile bracts at the base of the spikelet.

glutinous Sticky or gluelike.

head Usually a dense inflorescence of sessile flowers on a short or broadened axis.

hirsute With stiff hairs.

hypanthium A floral tube.

hyphae The basic filamentous growth form of a fungus.

imbricate Overlapping.

indehiscent Not opening by splitting along regular lines, or not opening.

indurate Hard.

inferior ovary An ovary positioned below the calyx.

inflorescence A flower cluster.

internode Portion of a stem or culm between 2 nodes.

involucel A set of sepal-like bracts slightly below the flower.

involucre A whorl of bracts subtending a flower or group of flowers.

involute Rolled inward from the edges with the upper surface within.

keel A central dorsal ridge; e.g., the keel of a boat.

lanceolate Lance shaped; much longer than broad; edges curved along the broad portion.

latex A milky juice.

leaflet A single division of a compound leaf.

legume A dry, dehiscent fruit with 2 valves, seeds attached along one line of dehiscence; fruit of the Fabaceae.

lemma The lowermost bract and usually the larger of the 2 bracts of a grass floret.

lenticel A corky spot on young bark.

ligulate Strap shaped; e.g., ray florets of the Asteraceae.

ligule A hairy or membranous appendage on the inner surface of a grass leaf at junction of the sheath and blade.

linear Long and narrow with more-or-less parallel sides; resembling a line.

lobe A segment of an organ; a division to about the middle.

loment A flat legume that is constricted between the seeds, breaking apart into 1-seeded segments.

lyrate Pinnatifid with the terminal lobe larger than the others.

m Meter; 100 centimeters.

margin The edge of a leaf blade.

monadelphous Having stamens that are united into one group by their filaments; occurs in members of the Malvaceae.

monoecious Having pistillate and staminate flowers on the same plant.

mucronate Tipped with a short spine.

mycelium A mass of fungal hyphae that forms an intertwined network.

node A point on a stem that bears a leaf or leaves.

nutlet A small, dry, nutlike fruit or seed.

obcordate Inversely cordate with the notch at the apex.

oblanceolate Inversely lanceolate.

obovate Inversely ovate.

obtuse Blunt or rounded at the apex.

ocrea A sheath around the stem at a node formed by the fusion of 2 stipules.

odd-pinnately compound With a terminal leaflet.

opposite Having leaves in pairs; on the opposite sides of the stem.

ovary Ovule-bearing, basal portion of a pistil.

ovary inferior Ovary positioned below the calyx.

ovary perigynous Stamens and petals inserted on the floral tube, i.e., borne around the ovary.

ovary superior The ovary located above the calyx and corolla.

ovate Egg shaped; the broadest portion below the middle.

palea The uppermost of the 2 bracts in a grass floret; usually smaller than the lemma.

palmate venation Divided from a common point; like the fingers of a hand.

panicle An indeterminate branching raceme; an inflorescence that is branched and re-branched.

paniculate Borne in a panicle.

papilionaceous Pertaining to an irregular corolla of some members of the Fabaceae; zygomorphic.

papilla A short, rounded projection or bump.

papillose With minute papillae.

pappus Modified, scalelike, or bristly calyx of the Asteraceae.

pedicel The stalk of a single flower.

peduncle The stalk of a cluster of flowers.

perennial A plant that continues to live for a number of years.

perianth The calyx and corolla of a flower.

perigynous With sepals, petals, and stamens attached to a calyx tube; surrounding but not attached to a superior ovary.

petal One of the inner leaflike parts of the flower, usually brightly colored.

petaloid Having a sepal that resembles a petal.

petiole A leaf stalk.

petiolule The stalk of a leaflet.

phyllary An involucral bract in the Asteraceae.

pilose With soft, relatively long hairs.

pinnate Branching on opposite sides of an axis.

pinnately compound Pertaining to a compound leaf with the leaflets arranged on each side of a petiole that is extended into a rachis.

pinnatifid Pinnately cleft into narrow lobes not reaching to the midrib.

pinnatisect Pinnately lobed to the midvein.

pistil The female reproductive structure of a flower; stigma, style, and ovary.

pistillate A flower with 1 or more pistils but no functional stamens; refers to a female flower or plant.

pollinium A pollen mass in the Asclepiadaceae and Orchidaceae.

poricidal Opening by pores.

prostrate Lying flat on the ground.

pubescence With short, soft hairs.

pubescent Covered with soft hairs.

pulvinus The swollen base of a petiole.

raceme A simple, elongated inflorescence with stalked flowers; the ordering of flowering from the base to the apex.

racemose In the shape of a raceme.

rachilla The axis that bears florets in a grass spikelet.

rachis An axis that bears flowers or leaflets.

radiate head With both ray and disk florets in a head; e.g., some members of the Asteraceae.

ray floret A strap-shaped or ligulate flower in the Asteraceae.

receptacle Where the florets are attached in the head of the Asteraceae.

recurved Curved downward or backward.

repand Wavy.

retrorse Bent downward or backward.

rhizome A creeping, underground stem bearing scalelike leaves.

rosette An arrangement of leaves radiating from a center, usually close to the ground.

rugose Rough.

scabrous Rough to the touch, like sandpaper.

scandent Climbing.

scape A flowering stem that is essentially leafless.

scarious Thin, dry, membranous, but not green.

schizocarp A dry fruit that splits into 1-seeded segments.

scorpioid An inflorescence that is coiled in the bud stage; like a scorpion's tail.

scurfy With scalelike particles that resemble dandruff.

sepal One of the outer leaflike parts of a flower, usually green but sometimes colored.

septate Divided by partitions.

serrate Having a saw-toothed leaf margin.

sessile Not stalked.

sheath A tubular structure surrounding a plant organ; the tubular, basal portion of a grass leaf.

simple Pertaining to a leaf with 1 flat green part.

sinuate With a wavy margin.

spathe A broad sheathing bract.

spatulate Oblong with an attenuated base.

spermatogonium (spermatogonial) Fruiting body of a rust fungus that results from the infection of a basidiospore.

spicate Resembling a spike.

spikelet A basic unit of a grass inflorescence, consisting usually of 2 glumes and 1 or more florets.

sporocarp A small, nutlike structure in some species of ferns, including giant salvinia, that contains megaspores and microspores, i.e., a heterosporous fern.

stamen Male reproductive structure of a flower; the filament and anther.

staminate Bearing stamens; refers to a male flower or a male plant.

staminode A sterile, non-pollen-producing stamen.

stellate Star shaped.

sterile Infertile.

stigma Portion of a pistil that receives pollen.

stipe A stalk below the ovary.

stipules A pair of appendages that are sometimes present at the point of attachment of the leaf to the stem.

style Portion of a pistil between the stigma and ovary.

sub- Prefix meaning slightly, almost, under, or almost.

subtend Standing below or close to; refers to a bract at the base of a flower.

succulent With thick, juicy parts.

superior ovary An ovary with sepals, petals, and stamens attached at or near the base of the ovary.

taproot The main root axis from which smaller roots arise.

tendril A segment of a stem or leaf modified into a slender, twining holdfast.

tepal A segment of a perianth that is not differentiated into a calyx and corolla.

tomentose Densely woolly.

trifoliolate Pertaining to a compound leaf bearing three leaflets.

truncate Ending abruptly; cut almost squarely at the end.

tubercle A small nodule.

umbel An inflorescence with the pedicels arising at about the same point.

umbellate Resembling an umbel; a flat-topped inflorescence with pedicels that arise from a common point.

undulate Pertaining to a wavy margin.

unisexual Of one sex; pistillate only or staminate only.

united Joined together.

urceolate Urn shaped.

utricle A small, 1-seeded, inflated fruit.

vein Conducting tissue; a vascular bundle.

viscid Sticky.

whorled With 3 or more branches at a node.

zygomorphic With or approaching bilateral symmetry.

LITERATURE CITED

Alderman, S. C., R. R. Halse, and J. F. White. 2004. A reevaluation of the host range and geographical distribution of *Claviceps* species in the United States. *Plant Disease* 88:63–81.

Alvarez, M. G. 1976. Primer catalogo de enfermedades de plantas mexicanas. *Fitofilo* 71:1–169.

Anonymous. 1971. *Common weeds of the United States*. Washington, DC: U.S. Department of Agriculture. Repr., Mineola, NY: Dover Publications.

Anonymous. 2001. *Aquatic vegetation in Texas: A guidance document*. Austin: Texas Parks and Wildlife Department.

Balesdent, M. H., M. Jedryczka, L. Jain, E. Mendes-Pereira, J. Bertrandy, and T. Rouxel. 1998. Conidia as a substrate for internal transcribed spacer-based PCR identification of members of the *Leptosphaeria maculans* species complex. *Phytopathology* 88:1210–17.

Barrett, S. C. H. 1989. Waterweed invasions. *Scientific American* 261:90–97.

Bell, G. P. 1997. Ecology and management of *Arundo donax* and approaches to riparian habitat restoration in southern California. In *Plant invasions: Studies from North America and Europe,* ed. J. H. Brock, M. Wade, P. Pysek, and D. Green, 103–13. Leiden, Netherlands: Backhuys Publishers.

Brown, L. E., and I. S. Elsik. 2002. Notes on the flora of Texas with additions and significant records. II. *Sida* 20:437–44.

Browning, M., L. Rowley, P. Zeng, J. M. Chandlee, and N. Jackson. 1999. Morphological, pathogenic, and genetic comparisons of *Colletotrichum graminicola* isolates from Poaceae. *Plant Disease* 83:286–92.

Byrd, J. D. 2003. Report of the 2002 cotton weed loss committee. *Proceedings of the 2002 Beltwide Cotton Conference.* Memphis, TN. CD-ROM.

Cook, C. G., and J. L. Riggs. 1993. Isolation and identification of *Phymatotrichum omnivorum* from kenaf in the Lower Rio Grande Valley of Texas. *Plant Disease* 77:1263.

Cordo, H. A., J. C. Deloach, and R. Ferner. 1981. Biological studies on two weevils, *Ochatia bruchi* and *Ochatina cretatus,* collected from *Pistia* and other aquatic plants in Argentina. *Annals of the Entomological Society of America* 74:363–69.

Cowan, H. B., ed. 1962. *Diseases of turfgrasses.* New York: Reinhold Publishing.

Crawford, J. L. 1970. Reduction in yield of cotton caused by parasitic diseases in 1969. *Plant Disease Reporter* 54:324.

Cummins, G. B. 1964. Uredinales of the Big Bend National Park and adjacent areas of Texas. *Southwestern Naturalist* 8:181–95.

Cummins, G. B., and H. C. Greene. 1966. A review of the grass rust fungi that have uredial paraphyses and aecia on *Berberis-Mahonia. Mycologia* 58:702–21.

Dahlberg, J. A. 2000. Classification and characterization of sorghum. In *Sorghum: Origin, history, technology, and production,* ed. C. W. Smith and R. A. Frederiksen, 99–130. New York: Wiley.

Davis, R. M., C. M. Heald, and L. W. Timmer. 1982. Chemical control of the citrus nematode on grapefruit. *Journal of the Rio Grande Valley Horticultural Society* 35:59–61.

Dean, J. L. 1966. Local infection of sorghum by the johnsongrass loose kernel smut fungus. *Phytopathology* 56:1342–44.

Deloach, C. J., J. C. Deloach, and H. A. Cordo. 1976. *Neohydronomous pulchellus,* a weevil attacking *Pistia stratiotes* in South America: Biology and host specificity. *Annals of the Entomological Society of America* 69:830–34.

DiTomaso, J. M., and E. A. Healy. 2003. *Aquatic and riparian weeds of the West.* Publication 3421. Davis: University of California, Agriculture and Natural Resources.

Dougherty, E. 1869. *Availability of the counties of Cameron and Hidalgo on the Lower Rio Grande* for agricultural, stock raising, and manufacturing purposes. Brownsville, TX: Sentinel Book and Job Printing Office.

Drees, B. M., and J. A. Jackman. 1998. *A field guide to common Texas insects*. Houston, TX: Gulf Publishing.

Dudley, T. L. 2000. *Arundo donax*. In *Invasive plants of California wildlands*, ed. C. C. Bossard, J. M. Randal, and M. C. Hosovsky, 53–58. Berkeley: University of California Press.

Dujovny, G., T. Usugi, and K. Shohara. 1998. Characterization of a potyvirus infecting sunflower in Argentina. *Plant Disease* 82:470–74.

Everitt, J. H., D. L. Drawe, and R. I. Lonard. 2002. *Trees, shrubs, and cacti of South Texas*. Rev. ed. Lubbock: Texas Tech University Press.

Farr, D., G. F. Bills, G. P. Chamuris, and A. Y. Rossman. 1989. *Fungi on plants and plant products in the United States*. St. Paul, MN: APS Press.

Fischer, G. W. 1953. *Manual of the North American smut fungi*. New York: Ronald Press.

Flores, D., J. H. Everitt, and J. W. Carlson. 2006. Assessing biological control damage of giant salvinia using remote sensing technologies. *Proceedings of the 20th Biennial Workshop Aerial Photography, Videography, and High Resolution Digital Imagery for Resource Assessment*. Bethesda, MD: American Society for Photogrammetry and Remote Sensing. CD-ROM.

Frederiksen, R. A., and G. N. Odvody, eds. 2000. *Compendium of sorghum diseases*. 2nd ed. St. Paul, MN: APS Press.

French, J. V., R. I. Lonard, and J. H. Everitt. 2003. *Cissus sicyoides* C. Linnaeus (Vitaceae), a potential exotic pest in the Lower Rio Grande Valley, Texas. *Subtropical Plant Science* 55:72–74.

Futrell, M. C., and R. A. Frederiksen. 1970. Distribution of sorghum downy mildew (*Sclerospora sorghi*) in the U.S.A. *Plant Disease Reporter* 54:311–14.

Gibb, T. J., and W. Buhler. 1995. Status of turfgrass insect and mite pests in the United States. In *Handbook of turfgrass insect pests*, ed. R. L. Brandenburg and M. G. Villani, 7–8. Lanham, MD: Entomological Society of America.

Godfrey, G. H. 1952. Foliage diseases of cucurbits in the Lower Rio Grande Valley of Texas in 1951. *Plant Disease Reporter* 36:69.

Goeden, R. D., and R. L. Kirkland. 1978. An insecticidal-check study of the biological control of puncturevine (*Tribulus terrestris* L.) by imported weevils, *Microlarinus lareynii* and *M. lypriformis* (Col.: Curculionidae). *Environmental Entomology* 7:349–54.

Graham, H. M., and O. T. Robertson. 1970. Host plants of *Heliothis virescens* and *H. zea* (Lepidoptera: Noctuidae) in the Lower Rio Grande Valley, Texas. *Annals of the Entomological Society of America* 63:1261–65.

Greenberg, S. M., T. W. Sappington, B. C. Legaspi, T. X. Liu, and M. Setamou. 2001. Feeding and life history of *Spodoptera exigua* (Lepidoptera: Noctuidae) on different host plants. *Annals of the Entomological Society of America* 94:566–75.

Groves, R. L., J. F. Walgenbach, J. W. Moyer, and G. G. Kennedy. 2001. Overwintering of *Frankliniella fusca* (Thysanoptera: Thripidae) on winter annual weeds with tomato spotted wilt virus and patterns of virus movement between susceptible weed hosts. *Phytopathology* 91:891–99.

———. 2002. The role of weed hosts and tobacco thrips, *Frankliniella fusca*, in the epidemiology of tomato spotted wilt virus. *Plant Disease* 86:573–82.

Gulya, T. J., P. J. Shiel, T. Freeman, R. L. Jordan, T. Isakeit, and P. H. Berger. 2002. Host range and characterization of sunflower mosaic virus. *Phytopathology* 92:694–702.

Hart, C. R., T. Garland, A. C. Barr, B. B. Carpenter, and J. C. Reagor. 2003. *Toxic plants of Texas: Integrated management strategies to prevent livestock losses*. College Station: Texas Cooperative Extension.

Hollowell, J. E., B. B. Shew, M. A. Cubeta, and J. W. Wilcut. 2003. Weed species as hosts for *Sclerotinia minor* in peanut fields. *Plant Disease* 87:197–99.

Holt, E. C. 1985. Buffelgrass: A brief history. In *Buffelgrass: Adaptation, management, and forage quality*, 1–5. MP-1575. College Station: Texas Agricultural Experiment Station.

Horne, C. W., ed and coord. 1988. *Texas plant diseases handbook*. Publication no. B-1140. College Station: Texas Agricultural Extension Service.

Horvath, B. J., and J. M. Vargas. 2004. Genetic variation among *Colletotrichum graminicola* isolates from four hosts using isozyme analysis. *Plant Disease* 88:402–6.

Idris, A. M., E. Hiebert, J. Bird, and J. K. Brown. 2003. Two newly described begomoviruses of *Macroptilium lathyroides* and common bean. *Phytopathology* 93:774–83.

Isakeit, T., G. N. Odvody, and R. A. Shelby. 1998. First report of sorghum ergot caused by *Claviceps africana* in the United States. *Plant Disease* 82:592.

Janick, J., R. W. Schery, F. W. Woods, and V. W. Rutton. 1981. *Plant science: An introduction to world crops.* San Francisco: W. H. Freeman.

Jones, S. D., and J. K. Wipff. 2003. *A 2003 updated checklist of the vascular plants of Texas.* Bryan, TX: Botanical Research Center. CD-ROM.

Kerenyi, Z., A. Moretti, A. Logrieco, and L. Hornok. 2000. Mating type diversity within *Gibberella fujikuroi* MP "D" (*Fusarium proliferatum*) assessed by PCR amplification of MAT-specific sequences. In *Proceedings of the 6th European Fusarium Seminar and 3rd COST 835 Workshop of Agriculturally Important Toxigenic Fungi,* ed. Helgard I. Nirenberg, Berlin, Germany, September 11–16.

Kingsbury, J. M. 1964. *Poisonous plants of the United States and Canada.* Englewood Cliffs, NJ: Prentice Hall.

Kirkland, R. L., and R. D. Goeden. 1978. Biology of *Microlarinus lareynii* (Col.: Curculionidae) on puncturevine in southern California. *Annals of the Entomological Society of America* 71:13–18.

Koike, H., and A. G. Gillaspie, Jr. 1989. Mosaic. In *Diseases of sugarcane: Major diseases,* ed. C. Richaud, B. T. Egan, A. G. Gillaspie, Jr., and C. G. Hughes, 301–2. Amsterdam, Netherlands: Elsevier Science Publishers.

Langeland, K. A. 1996. *Hydrilla verticillata* (L. F.) Royle (Hydrocharitaceae): "The perfect aquatic weed." *Castanea* 612:293–304.

Li, R., S. Salih, and S. Hurtt. 2004. Detection of geminiviruses in sweetpotato by polymerase chain reaction. *Plant Disease* 88:1347–51.

Lonard, R. I. 1993. *Guide to grasses of the Lower Rio Grande Valley, Texas.* Edinburg: University of Texas–Pan American Press.

Lonard, R. I., and F. W. Judd. 2002. Riparian vegetation of the Lower Rio Grande. *Southwestern Naturalist* 47:420–32.

Lonard, R. I., A. T. Richardson, and N. L. Richard. 2004. The vascular flora of the Palo Alto National Battlefield Historic Site, Cameron County, Texas. *Texas Journal of Science* 56:15–34.

Lopes, S. A., S. Marcussi, S. C. Z. Torres, V. Sluza, C. Fagan, S. C. Franca, N. G. Fernandes, and J. R. S. Lopes. 2002. Weeds as alternative hosts of the citrus, coffee, and plum strains of *Xylella fastidiosa* in Brazil. *Plant Disease* 87:544–49.

Marley, P. 1995. *Cynodon dactylon:* An alternative host for *Sporisorium sorghi,* the causal organism of sorghum covered smut. *Crop Protection* 14:491–93.

Martin, E. C., A. M. Baltazar, J. M. Ramos, S. K. De Datta, L. T. Kok, and E. G. Rajotte. 2002. Efficacy of *Spoladea recurvalis* as biological control agent against *Trianthema portulacastrum* L. Annual report to the Integrated Pest Management CRSP, Blacksburg, VA.

Mayeux, H. S., Jr., C. J. Scifres, and R. A. Crane. 1980. *Ericameria austrotexana* and associated range forage responses to herbicides. *Weed Science* 28:602–6.

Milhail, J. D. 1992. *Macrophomina phaseolina.* In *Methods for research on soilborne phytopathogenic fungi,* ed. L. L. Singleton, J. D. Milhail, and C. M. Rush. St. Paul, MN: APS Press.

Mitchell, D. S. 1976. The growth and management of *Eichhornia crassipes* and *Salvinia* spp. in their native environment and in alien situations. In *Aquatic weeds in Southeast Asia,* ed. C. K. Varshncy and J. Rzoska, 167–75. The Hague, Netherlands: Dr. W. Junk Publisher.

Monaco, T. J., S. C. Weller, and F. M. Ashton. 2002. *Weed science: Principles and practice.* 4th ed. New York: Wiley.

Nichols, S. A., and B. H. Shaw. 1986. Ecological life histories of three aquatic nuisance plants, *Myriophyllum spicatum, Potamogeton crispus,* and *Elodea canadensis. Hydrobiologia* 131:3–21.

Odvody, G. N., and D. B. Madden. 1984. Leaf sheath blights of *Sorghum bicolor* caused by *Sclerotium rolfsii* and *Gloeocercospora sorghi* in South Texas. *Phytopathology* 74:264–68.

O'Neill, N. R., and G. R. Bauchan. 2000. Sources of resistance to anthracnose in the annual *Medicago* core collection. *Plant Disease* 84:261–67.

———. 2003. Reactions in the annual *Medicago* spp. core germ plasm collection to *Phoma medicaginis. Plant Disease* 87:557–62.

Orellana, R. G. 1971. Fusarium wilt of sunflower, *Helianthus annuus:* First report. *Plant Disease Reporter* 55:1124–25.

Perdue, R. E., Jr. 1958. *Arundo donax:* A source of musical reeds and industrial cellulose. *Economic Botany* 12:368–404.

Pest Management Recommendations for Field Crops. 2003. http://www.agnr.umd.cdu/MCE/Publications/PDFs/EB237-NED237_4–9.pdf (accessed April 2005).

Pimentel, D., L. Lach, R. Zuniga, and D. Morrison. 2000. Environmental and economic costs of nonindigenous species in the United States. *Bioscience* 50:53–65.

Rhodes, L. H., primary collator. 2001. Common names of plant diseases: Diseases of alfalfa (*Medicago sativa* L.). American Phytopathological Society. http://www.apsnet.org/online/common/names/alfalfa.asp (accessed June 2005).

Robinson, A. F., C. M. Heald, S. L. Flanagan, W. H. Thames, and J. Amador. 1987. Geographical distribution of *Rotylenchulus reniformis, Meloidogyne incognita,* and *Tylenchulus semipenetrans* in the Lower Rio Grande Valley as related to soil texture and land use. *Annals of Applied Nematology* 1:20–25.

Rodriguez-Alvarado, G., S. Fernandez-Pavia, R. Creamer, and C. Liddell. 2002. Pepper mottle virus causing disease in chile peppers in southern New Mexico. *Plant Disease* 86:603–5.

Rogers, C. E., T. E. Thompson, and D. E. Zimmer. 1978. Rhizopus head rot of sunflower: Etiology and severity in the southern plains. *Plant Disease Reporter* 62:769–71.

Romberg, M. K., J. J. Nujnez, and J. J. Farrar. 2004. First report of powdery mildew on potato caused by *Golovinomyces cichoracearum* in California. *Plant Disease* 88:309.

Rosewich, U. L., R. E. Pettway, B. A. McDonald, R. R. Duncan, and R. A. Frederiksen. 1998. Genetic structure and temporal dynamics of a *Colletotrichum graminicola* population in a sorghum disease nursery. *Phytopathology* 88:1087–93.

Rummel, D. R., and M. D. Arnold. 1992. Status of the puncturevine seed weevil in the Texas Southern High Plains. *Southwestern Entomologist* 17:347–49.

Rush, C. M., T. J. Gerik, and C. M. Kenerley. 1985. Atypical disease symptoms associated with Phymatotrichum root rot of cotton. *Plant Disease* 69:534–37.

Schrock, D., ed. 2004. *Home gardener's problem solver.* Des Moines, IA: Meredith Books.

Shurtleff, M. C. 1980. *Compendium of corn diseases.* 2nd ed. St. Paul, MN: American Phytopathological Society.

Smiley, R. W., P. H. Dernoeden, and B. B. Clarke. 2005. *Compendium of turfgrass diseases.* 3rd ed. St. Paul, MN: APS Press.

Smith, C., ed. 1993. *The Ortho home gardener's problem solver.* San Ramon, CA: Ortho Books.

Smither-Kopperl, M. L., R. Charudattan, and R. D. Berger. 1998a. Dispersal of spores of *Fusarium culmorum* in aquatic systems. *Phytopathology* 88:382–88.

———. 1998b. *Plectosporium tabacinum,* a pathogen of the invasive aquatic weed *Hydrilla verticillata* in Florida. *Plant Disease* 83:24–28.

Sperry, O. E., J. W. Dollahite, G. O. Hoffman, and B. J. Camp. 1968. *Texas plants poisonous to livestock.* B-1028. College Station: Texas Agricultural Experiment Station.

Sprague, R. 1950. *Diseases of cereals and grasses in North America.* New York: Ronald Press.

Spring, O., H. Voglmayr, A. Riethmuller, and F. Oberwinkler. 2003. Characterization of a *Plasmopara* isolate from *Helianthus* × *laetiflorus* based on cross infection, morphological, fatty acids and molecular phylogenetic data. *Mycological Progress* 2:163–70.

Stoddard, A. A. 1989. The phytogeography and paleofloristics of *Pistia stratiotes* L. *Aquatics* 11:21–24.

Stutzenbaker, C. D. 1999. *Aquatic and wetland plants of the western Gulf Coast.* Austin: Texas Parks and Wildlife Press.

Subcommittee on Standardization of Common and Botanical Names of Weeds. 1966. Composite list of weeds. *Weeds* 14:347–86.

Towers, G. H. N. 1979. Contact hypersensitivity and photodermatitis evoked by Compositae. In *Toxic plants,* ed. A. D. Kinghorn, 171–82. New York: Columbia University Press.

Tucker, R. W. E. 1940. An account of *Diatraea saccharalis* F. with special references to its occurrence in Barbados. *Tropical Agriculture* 17:133–38.

Westphal, A., and J. R. Smart. 2003. Depth distribution of *Rotylenchulus reniformis* under different tillage and crop sequence systems. *Phytopathology* 93:1182–89.

Wilson, J. P. 1999. *Pearl millet diseases: A compilation of information on the known pathogens of pearl millet,* Pennisetum glaucum *(L.) R. Br.* Agriculture Handbook No. 716. Washington, DC: U.S. Department of Agriculture, Agricultural Research Service.

Yaege, J. R., and D. L. Stuteville. 2000. Reactions in the annual *Medicago* core germplasm collection to two isolates of *Peronospora trifoliorum* from alfalfa. *Plant Disease* 84:521–24.

———. 2002. Reactions of accessions in the annual *Medicago* core germplasm collection of *Erysiphe pisi. Plant Disease* 86:312–15.

Yang, S., R. W. Berry, E. S. Luttrell, and T. Vongkaysone. 1984. A new sunflower disease in Texas caused by *Diaporthe helianthi. Plant Disease* 68:254–55.

Yang, S. M., J. B. Morris, P. W. Unger, and T. E. Thompson. 1979. Rhizopus head rot of cultivated sunflower in Texas. *Plant Disease Reporter* 63:833–35.

Yang, S. M., and D. F. Owen. 1982. Symptomatology and detection of *Macrophomina phaseolina* in sunflower plants parasitized by *Cylindrocopturus adspersus* larvae. *Phytopathology* 72:819–21.

Zhang, W. M., and A. K. Watson. 1997. Effect of dew period and temperature on the ability of *Exserohilum monoceras* to cause seedling mortality of *Echinochloa* species. *Plant Disease* 81:629–34.

Other References

Almaraz, T., and G. Durrieu. 1997. Ustilaginales from the Spanish Pyrenees and Andorra. *Mycotaxon* 65:223–36.

Anonymous. 1996. *Corn insect identification guide.* Wayne, NJ: American Cyanamid Company.

Browning, H., and M. Hussey. 1987. Susceptibility of "Tifton 68" and "Coastal" bermudagrass to the Mexican rice borer. *Crop Science* 27:358–60.

Cobb, P., and T. Mack. 1989. A rating system for evaluating tawny mole cricket, *Scapteriscus vicinus* Scudder, damage (Orthoptera: Gryllotalpidae). *Journal of Entomological Science* 24:142–44.

Datnoff, L., M. Elliott, and J. Krausz. 1997. Cross pathogenicity of *Gaeumannomyces graminis* var. *graminis* from bermudagrass, St. Augustinegrass, and rice in Florida and Texas. *Plant Disease* 81:1127–31.

Dickson, J. G. 1956. *Diseases of field crops.* New York: McGraw-Hill.

Esele, J. P. 1995. Foliar and head diseases of sorghum. *African Crop Science Journal* 3:185–89.

Grisham, M. P. 2000. Mosaic. In *A guide to sugarcane diseases,* ed. P. Rott, R. Bailey, J. C. Comstock, B. Croft, and S. Saumtally, 249–54. Montpellier, France: CIRAD/ISSCT.

Harding, J. A. 1976. *Heliothis* spp.: Seasonal occurrence, hosts and host importance in the Lower Rio Grande Valley. *Environmental Entomology* 5:666–68.

Hurd, B., and M. P. Grisham. 1983. *Rhizoctonia* spp. associated with brown patch of Saint Augustinegrass. *Phytopathology* 73:1661–65.

Johnson-Cicalese, J., and C. Funk. 1990. Additional host plants of four species of billbug found on New Jersey turfgrasses. *Journal of the American Society for Horticultural Science* 115:608–11.

Jorda, C., A. Ortega, and M. Juarez. 1995. New hosts of tomato spotted wilt virus. *Plant Disease* 79:538.

Kumar, S., and G. Singh. 1997. Efficacy and selectivity of tralkoxydim alone or in combination with isoproturon in wheat (*Triticum aestivum*) streak virus, a previously undescribed plant rhabdovirus infecting Bermuda grass and maize in the Mediterranean area. *Phytopathology* 75:1094–98.

Lockhart, B., N. Khaless, L. Lennon, and M. El Maatauoi. 1985. Properties of Bermudagrass etched-line virus, a new leafhopper-transmitted virus related to maize rayado fino and oat blue dwarf viruses. *Phytopathology* 75:1258–62.

Mohamed, M., S. Quisenberry, and D. Moellenbeck. 1992. 6,10,14-Trimethylpentadecan-2-one: A Bermuda grass phagostimulant to fall armyworm (Lepidoptera:Noctuidae). *Journal of Chemical Ecology* 18:673–82.

Pashley, D. S., S. Quisenberry, and T. Jamjanya. 1987. Impact of fall armyworm (Lepidoptera: Noctuidae) host strains on the evaluation of Bermuda grass resistance. *Journal of Economic Entomology* 80:1127–30.

Pratt, R. G. 2003. First report of infection of Bermudagrass by *Bipolaris sorokiniana* in the southeastern United States. *Plant Disease* 87:1265.

Sharma, R., and S. Sachan. 1994. New hosts of rust fungi from Himachal Pradesh. *Advances in Plant Sciences* 7:154–58.

Tisserat, N., S. Hulbert, and A. Nus. 1991. Identification of *Leptosphaeria korrae* by clone DNA probes. *Phytopathology* 81:917–21.

Viji, G., W. Uddin, N. R. O'Neill, S. Mischke, and J. A. Saunders. 2004. Genetic diversity of *Sclerotinia homeocarpa* isolates from turfgrasses from various regions of North America. *Plant Disease* 88:1269–76.

Wu, W. S., and V. P. Wang. 1994. First report of the occurrence of *Bipolaris maydis* on bermudagrass. *Plant Disease* 78:926.

Yamada, W., T. Shiomi, and H. Yamamoto. 1956. Studies on the stripe disease of rice plant, 3: Host plants, incubation period in the rice plant, and retention of over-wintering of the virus in the insect, *Delphacodes striatella* Fallen. Special Bulletin, Okayama Prefectural Agriculture Experiment Station 55:35–36.

Zuniga, G. E., V. H. Argandona, H. M. Niemeyer, and L. J. Corcuera. 1983. Hydroxamic content in wild and cultivated Gramineae. *Phytochemistry* 22:2665–68.

INDEX